AF360826

Grid-to-Vehicle (G2V) and Vehicle-to-Grid (V2G) Technologies

Grid-to-Vehicle (G2V) and Vehicle-to-Grid (V2G) Technologies

Editors

Sekyung Han
Moses Amoasi Acquah

MDPI • Basel • Beijing • Wuhan • Barcelona • Belgrade • Manchester • Tokyo • Cluj • Tianjin

Editors
Sekyung Han
Kyungpook National University
Korea

Moses Amoasi Acquah
Keimyung University
Korea

Editorial Office
MDPI
St. Alban-Anlage 66
4052 Basel, Switzerland

This is a reprint of articles from the Special Issue published online in the open access journal *Energies* (ISSN 1996-1073) (available at: https://www.mdpi.com/journal/energies/special_issues/ G2V_V2G).

For citation purposes, cite each article independently as indicated on the article page online and as indicated below:

LastName, A.A.; LastName, B.B.; LastName, C.C. Article Title. *Journal Name* **Year**, *Article Number*, Page Range.

ISBN 978-3-03943-444-2 (Hbk)
ISBN 978-3-03943-445-9 (PDF)

Contents

About the Editors

Sekyung Han (Professor). Prof. Han is Associate Professor at the Department of Electrical Engineering, Kyungpook National University. He received his Ph.D. degree in Information Science and Technology at the University of Tokyo, Japan, and Master degree in Electrical Engineering and Computer Science, Seoul National University, South Korea, in 2012 and 2007, respectively. Prof. Han has been systematically reviewing, writing, speaking, and providing solutions on the topic of optimization and cost effectiveness of smart grid, battery, and electric vehicles since 2012. He serves as a council member of Gerson Lehrman Group, Expert in Battery Systems, and a board member at Daegu International Future Auto Expo(DiFa) in addition to being Editor of the *Energies* Special Issue "Grid-to-Vehicle (G2V) and Vehicle-to-Grid (V2G) Technologies". Prof. Han's primary interests are teaching, investigation, and developing solutions in smart grid, energy management system, battery management system (BMS), and electric vehicles.

Moses Amoasi Acquah (Professor). Prof. Acquah is Assistant Professor at the Department of Electrical Energy Department Keimyung University, South Korea. He received his Ph.D. degree in Electrical Engineering at Kyungpook National University, South Korea, and MPhil degree in Computer Engineering from the University of Ghana, Legon, in 2014 and 2018, respectively. Prof. Acquah served as co-Editor for the *Energies* Special Issue "Grid-to-Vehicle (G2V) and Vehicle-to-Grid (V2G) Technologies" and a member of the Institute of Electrical Engineers (IEEE), Power and Energy Society. His main interests include teaching, research, and developing solutions in vehicle-to-grid (V2G), energy management system, big data, and machine learning and optimization in smart grid.

Preface to "Grid-to-Vehicle (G2V) and Vehicle-to-Grid (V2G) Technologies"

The present volume contains proceedings of the Special Issue: "Grid-to-Vehicle (G2V) and Vehicle-to-Grid (V2G) Technologies", submitted from 1 May 2019 to 31 May 2020.

In recent years, there has been a global outcry for the decarbonization of energy systems to provide a secure and sustainable energy supply while cutting down on greenhouse gas emissions (GHGs) in an effort to avoid global warming. With the explosion in use of battery energy storage, electric vehicles (EV) are considered an efficient eco-friendly means of transportation. They also play a role through their interaction with electricity grids, delivering power as well as controlling the charging rate for a faster charging time. EVs are able to meet this role due to grid-to-vehicle (G2V) and vehicle-to-grid (V2G) operation, providing bidirectional power flow to tackle the twin challenges of faster charging and providing ancillary services to the grid. This volume contains articles on "Grid-to-Vehicle (G2V) and Vehicle-to-Grid (V2G) Technologies" such as solutions in G2V and V2G, including but not limited to the operation and control of gridable vehicles, energy storage and management systems, charging infrastructure and chargers, load forecasting for distributed systems, V2G interfaces, communication standards, and charging topologies as well as environmental impacts and economic benefits.

Special thanks to all who kindly contributed their papers for this issue: the esteemed reviewers and the editors of MDPI *Energies* for their kind help and co-operation. We are also indebted to the MDPI *Energies* Editorial Office. The editorial assistance of the publishing and production teams at *Energies* is acknowledged for the prompt processing and production of work of this Special Issue that successfully kept production on schedule.

Sekyung Han, Moses Amoasi Acquah
Editors

Article

An Ensemble Stochastic Forecasting Framework for Variable Distributed Demand Loads

Kofi Afrifa Agyeman, Gyeonggak Kim, Hoonyeon Jo, Seunghyeon Park and Sekyung Han *

Electrical Engineering Department, Kyungpook National University, Daegu 41566, Korea;
kofiagyeman@knu.ac.kr (K.A.A.); kakkyoung2@gmail.com (G.K.); hyjo1006@naver.com (H.J.);
mrbbbark@gmail.com (S.P.)
* Correspondence: skhan@knu.ac.kr

Received: 15 April 2020; Accepted: 21 May 2020; Published: 25 May 2020

Abstract: Accurate forecasting of demand load is momentous for the efficient economic dispatch of generating units with enormous economic and reliability implications. However, with the high integration levels of grid-tie generations, the precariousness in demand load forecasts is unreliable. This paper proposes a data-driven stochastic ensemble model framework for short-term and long-term demand load forecasts. Our proposed framework reduces uncertainties in the load forecast by fusing homogenous models that capture the dynamics in load state characteristics and exploit model diversities for accurate prediction. The ensemble model caters for factors such as meteorological and exogenous variables that affect load prediction accuracy with adaptable, scalable algorithms that consider weather conditions, load features, and state characteristics of the load. We defined a heuristic trained combiner model and an error correction model to estimate the contributions and compensate for forecast errors of each prediction model, respectively. Acquired data from the Korean Electric Power Company (KEPCO), and building data from the Korea Research Institute, together with testbed datasets, were used to evaluate the developed framework. The results obtained prove the efficacy of the proposed model for demand load forecasting.

Keywords: Bayesian; deep neural network; demand load forecast; distributed load; ensemble algorithm stochastic; K-means

---

1. Introduction

Developments in the invasive use of grid-flexibility options, such as demand-side management (DSM), require pliability in load prediction mechanisms to match temporal and spatial differences between energy demand and supply [1]. Recent advancements in DSMs, such as vehicle-to-grid (V2G) technologies and renewable energy policies, induce new perspectives for energy demand-supply imbalance management [2]. A critical factor in this is the reliable prediction mechanism of demand loads [3]. Thus, much attention has been given to demand load forecast mechanisms over the past decade [4]. A significant drawback of these prediction mechanisms is the lack of reliable high-sampled historical demand data, unscalable predictive models to match consumption patterns, and insufficient information on load prediction state uncertainties [5,6]. Energy consumption differs among load types, consumption behavior, and time of energy usage. For loads with routine energy consumption, such as office building loads, the energy sequence is stationary with minimal energy variations within the operation periods. However, for non-routine energy consumption or generation loads, such as hotels and renewable power, respectively, the energy sequence is randomized, with many variabilities in the energy profile patterns. Such a situation can pose a challenge to predict with generalized prediction models.

Demand load forecasting is an established yet still very active research area [7]. The literature on prediction models on energy consumption is enormous, with recent studies harnessing the power of machine learning (ML) to develop highly-generalized predictive models. Before this, classical predictive methods dwelled mainly on statistical analysis [8–10]. Consumption patterns were stable with fewer variations with these loads; hence, models such as support vector regression (SVR) and auto-regressive integrated moving average (ARIMA) were used for short-term prediction of loads [11–18]. The relevance of these methods is dependent on the extensive dataset with collinear measuring variables, which, in most cases, are difficult to come by. Therefore, classical predictive methods were, as a result, incapable of capturing random variations in the data patterns [19]. Efforts to improve the forecasting methods incorporating such diversity and unevenness prompted attempts to replace classical regression models with ML techniques [20]. Predominate among these techniques is the artificial neural network (ANN). The ANN is already known for its dominant utilization in energy forecasting methods [13,21,22]. However, because of its inherent complexity, it mostly leads to overfitting. So, an ensemble model that harnesses the merit of ANN potential, together with other ML algorithms, could be devised for high accuracy load prediction. Recently, Wang et al. proposed an ensemble forecasting method for the aggregated load with subprofile [23]. In this work, the load profile is clustered into subprofiles, and forecasting is conducted on each group profile. Apart from the fact that this algorithm is based on a fine-grained subprofile, which may not be readily accessible from every energy meter, cluster members with similar features but different load profiles are problematic to cluster. Hence, a centroid representation of a cluster may lead to higher variance in load prediction. In [24], Wang et al. proposed a combined probabilistic model for load forecast based on a constrained quantile regression averaging method. This method is based on an interval forecast instead of a point forecast. Apart from increasing computational time required for bootstrapping, much data is needed, and interval resolution may not be optimal for other data. Given this, this paper focuses on a predictive ensemble with limited available historical datasets to develop a scalable online predictive model for demand load forecasting. In this study, date meta-data parameters and weather condition variables serve as inputs to the proposed framework. As opposed to the models mentioned above, the proposed prediction model is not based on any specific data sampling interval or strict prediction interval. The model is adaptable to different distributed demand loads with a varying number of predictors at various sampling intervals. The contributions of this study are to:

1. Define an online stochastic predictive framework with a computation time of less than a minute.
2. Define a prediction model capable of training a robust forecast model with a single or limited historical dataset.
3. Define a prediction framework scalable and adaptable to different distributed demand load types.
4. Define an error correction model capable of compensating forecast error.

The remainder of the paper is organized as follows. Section 2 discusses some challenges in demand load forecasting. Section 3 describes the probabilistic load forecasting model generation for stochastic demand load forecasting and error compensation methods. In Section 4, the results of the parametric models and the ensemble forecasting model on different case studies are presented.

2. Challenges in Load Forecasting

Load forecasting is a technique adopted by power utilities to predict the energy needed to meet generation to maintain grid stability. Data-gathering methods used in such an exercise are often unreliable, sometimes resulting in missing, nonsensical, out-of-range, and NaN(i.e., Not a Number) values. The presence of irrelevant and redundant information or noisy and unreliable data can affect knowledge discovery during the model training phase. The accuracy of forecasting is of great significance for both the operational and managerial loading of a utility. Despite many existing load forecasting methods, there are still significant challenges regarding demand load forecast accuracy. These challenges are data integrity verification, adaptive predictive model design, forecast error compensation, and dynamic model selection issues.

2.1. Unreliable Data Acquisition

Demand load forecasting is based on expected weather condition information. Accurate weather forecast is difficult to achieve because of changes in weather conditions from sudden natural occurrences. Thus, the demand load forecast may thus differ due to the actual weather condition information. Apart from the weather feature parameters, other forecasting features, such as data meta-data and consumed load, are required for demand load forecasting. The acquisition of this dataset could be unreliable due to the break-down or malfunctioning of energy meters. The dataset may thus contain non-sequential, missing, or nonsensical data. The consequential effects of such an event render the dataset highly unreliable and not suitable for accurate load forecasting.

2.2. Adaptive Predictive Modeling

Demand load forecasting complexities are influenced by the nature of the demand load, which is a result of consumer behavior changes, energy policies, and load type. The behavioral changes of energy consumers result in different energy usage patterns. For instance, building load type is determined by the hosted activities: commercial, residential, or industrial. Energy consumption in commercial and industrial buildings occurs in the light of routine activities that are derived from either uniform equipment operations or the implied consistency of organized human activities [25]. Demand load is affected by both exogenous (i.e., weather conditions) and endogenous (i.e., type of day) parameters [26,27]. With a stationary demand load, such as an office building load, energy consumption follows a specific pattern; hence, variability in energy consumption is not volatile.

Conversely, non-stationary buildings such as hotels have a high-frequency fluctuation in their energy demand sequence because of randomized operation conditions, such as varying occupancy levels. The changing pattern leads to poor prediction accuracy with an unscalable predictive algorithm. To mitigate such problems, conventionally, for grid load forecast, utility operators use manual methods that rely on a thorough understanding of a wide range of contributing factors based on upcoming events or a particular dataset. Relying on manual forecasting is unsustainable due to the increasing number of complexities of the prediction. Hence, the predictive model for load should be adaptive to the changing conditions.

2.3. Transient-State Forecast Error

The estimated load forecast may change as a result of a sudden natural occurrences, such as tsunami, typhoons, or holidays. Such changes are temporary and may not occur often. Thus, the day-ahead predictive parameters may not reflect these sudden changes. Hence, the predictive model forecast may deviate from the expected forecast. The predictive model captures only steady-state parameters when the load is modeled to fit historical data. The steady-state predictive model ignores transient forecast errors. Such a phenomenon significantly reduces the accuracy of the load forecast.

2.4. Model Selection Criteria

All statistical forecasting methods could be used to fit a model on a dataset. However, it is difficult to fit a model that captures all the variabilities in a demand load dataset, considering the numerous complex factors that influence demand load forecasting. Additionally, obtaining accurate demand load forecasts based only on parameters such as weather information and other factors that influence consumption may not always be correct since, under certain circumstances, some predictive models have higher prediction accuracy than others. Hence, the criteria for selecting the best fit model under certain conditions is critical for accurate demand load forecasting. Given this, various methods are proposed in this framework to address these challenges, as mentioned earlier.

3. Probabilistic Load Forecasting Model Generation

In this study, we based our proposed framework on the ensemble forecasting method. The proposed forecasting framework, as shown in Figure 1, involves the process of demand consumption data and weather information acquisition, data integrity risk reduction, forecast error compensation, point-forecast, and stochastic forecast estimation. The ensemble forecasting method is an integration of a series of homogenous parametric models as a random forecast model [24]. The set of parallel parametric models train and forecast load independently. In this study, the forecasting process is in two stages: point-forecast model generation and ensemble processes. The point-forecast generation process involves mechanisms in generating deterministic forecasts with point-forecast models. This process comprises parametric model selection techniques and data integrity risk reduction methods. With the ensemble process, the model scheme defines strategies for optimal forecast model selection, forecast error compensation, and stochastic forecasts. Since parametric model selection is not the primary consideration of this study, we introduce only three typical parametric models. In the last decade, several parametric models have been selected for demand load predictive modeling. Notable among them are the artificial neural network (ANN), K-means, and Bayesian [28–39] approaches. The choice of these models for demand load prediction is because of their acute sensitivity to sequential data prediction. Our proposed framework is adaptable to multiple parametric or meta-parametric models for the efficient prediction of sequential demand load.

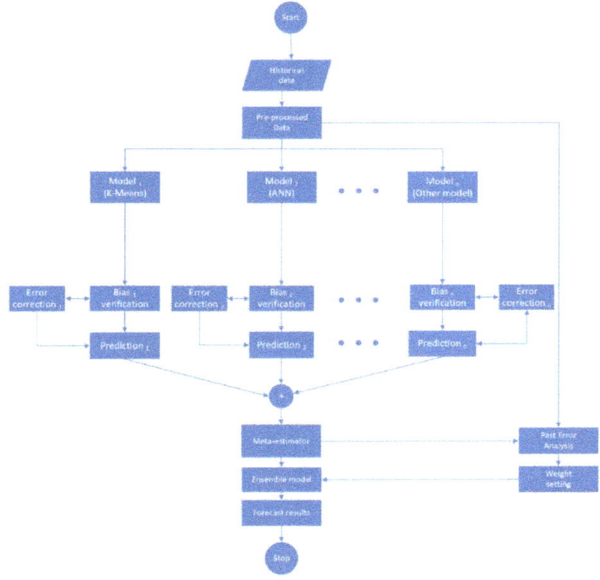

Figure 1. Demand load forecast flowchart.

3.1. Data Integrity Risk Reduction

As part of the study, preprocessing analysis was performed on the dataset to increase the integrity of the dataset. We defined a risk reduction method as part of this study to detect outliers and also compensate for irregular or missing data. The first stage of the risk reduction method deals with data outlier detection. The acquired demand load data contains outliers because of measurement latencies, changes in systems behavior, or fault in measuring devices. In this study, boxplot and generalized extreme studentized deviate (ESD) algorithms were adopted to detect irregularities in the demand load data. These methods are common statistical techniques used to identify hidden patterns and multiple outliers in data [40]. The technique depends on two parameters, α and r, being the probability of false outliers and the upper limit of expected number of potential outliers, respectively. As mentioned by Carey et al. [41], the maximum number of potential outliers in a dataset is defined by the inequality in Equation (1).

$$r < 0.5(n-1) \tag{1}$$

where n is the number of observations in the data set. With the upper bound of r defined, the generalized ESD test performs r separate tests: a test for one outlier, two outliers, and so forth up to r outliers.

$$X : \{x_1, x_2, x_3, \dots, x_n\} \tag{2}$$

For each r separate test, we compute the test statistic, R_i, that maximizes $|x_i - \bar{x}|$ for all observations as specified in Equation (3), where $\bar{x}$ and s denote the sample mean and the standard deviation, respectively. The observation x_i that maximises R_i is then removed and R_i recomputed with $(n-1)$ observations. This process is repeated for all r with the outcome test statistics of $R_1, R_2, \cdots, R_r$.

$$R_i = \frac{max|x_i - \bar{x}|}{s} \tag{3}$$

The critical value, λ_i, is therefore computed for each r test statistic in Equation (4) with tail area probability, p, defined in Equation (5), where $t_{p,n-i-1}$ is the $100p$ percentage point from the t-distribution with $(n-i-1)$ degrees of freedom. The pair of test statistics and critical values are arranged in descending order. The number of outliers is therefore determined by finding the largest i such that $R_i > \lambda_i$.

$$\lambda_i = \frac{(n-1)t_{p,n-i-1}}{\sqrt{\left(n-i-1+t^2_{p,n-i-1}\right)(n-i+1)}}; \ \forall \ i \in r \tag{4}$$

$$p = 1 - \left[\frac{\alpha}{2(n-i+1)}\right] \tag{5}$$

Following the outlier detection is the data imputation method. Missing data as a result of outlier removal or sampling error are compensated for with the data imputation method. Using demand load profiles with missing data can introduce a substantial amount of bias, rendering the analysis of the data more arduous and inefficient. The imputation model defined in this study preserves the integrity of the load profiles by replacing missing data with an estimated value based on other available information. The proposed reduction model in this study is in three stages of imputation: listwise deletion, hot-deck, and Lagrange regression model.

For training analysis, from a dataset X with feature variables M and number of observations N, the listwise deletion method deletes a set of records with missing data values of any of the feature variables. The listwise deletion method produces a reduced dataset, X^*, as defined in Equation (6).

$$X^* = \left[X^*_1, X^*_2, \cdots X^*_j\right]; \ \forall \ j \in N \tag{6}$$

$$X^*_j = X_j : \exists \ l_{i,j} \cap \exists \ l_{i+1,j} \cap \cdots \cap \exists \ l_{i+m,j} = 1 \tag{7}$$

$$l_{i,j} = \begin{cases} 1; & \exists \ x_{i,j} \forall \ i \in M \\ 0; & \nexists \ x_{i,j} \forall \ j \in N \end{cases} \tag{8}$$

This process does not cause bias, but decreases the power of the analysis by reducing the sufficient sample size. For continuous missing data values (i.e., >3, based on the sampling interval), the listwise method could introduce bias. Thus, for a single missing value, the hot-deck technique is adapted—the hot-deck procedure imputes missing values with previously available data of the same feature variable as defined in Equation (9). The hot-deck technique precedes the listwise deletion method.

$$X^* = \left[X_1^*,\ X_2^*, \cdots X_j^*\right];\ \forall\ j \in N \tag{9}$$

$$X_j^* = \left[X_{1,j}^*, X_{2,j}^*, \cdots X_{m,j}^*\right];\ \forall\ m \in M \tag{10}$$

$$x_{i,j}^* = \begin{cases} x_{i,j}; & \exists\ x_{i,j} \forall\ i \in M \\ x_{i,j-1}; & \nexists\ x_{i,j} \forall\ j \in N \end{cases} \tag{11}$$

For multiple data imputation, an ensemble method made up of regression and Lagrange imputation was adapted. There are seasonal, cyclic, and trend patterns in a demand load dataset. With a week-ahead daily load profile, as depicted in Figure 2, the similarity in the patterns of the two load profiles of the same season and day type is evident. Thus, from the immediate prior week, daily load profile information could be used to estimate missing data values in the week-ahead daily load profile.

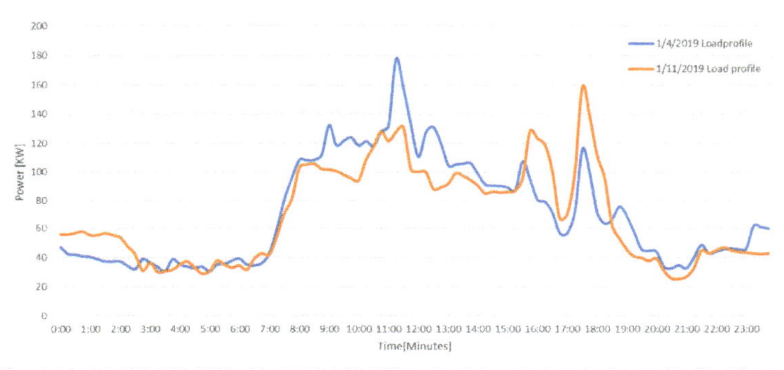

Figure 2. Week-ahead daily load profile.

To achieve this, we estimated the missing data values with a regression imputation method based on the observed values. However, the perfect prediction regression equation is impractical in some situations (e.g., with a limited number of observations). Hence, the standard error associated with the forecast can be reduced with the demand profile of the previous week. Thus, we estimated the deviation and the rate of deviation, R, of the previous demand load profile. From this, we defined a Lagrange imputation model with the load deviation rate as defined in Equation (12). The initial predicted missing values are modified with the Lagrange model, as shown in Figure 3.

$$y = \frac{(R-R_2)(R-R_3)\cdots(R-R_n)}{(R_1-R_2)(R_1-R_3)\cdots(R_1-R_n)}y_1 + \\ \frac{(R-R_1)(R-R_3)\cdots(R-R_n)}{(R_2-R_1)(R_2-R_3)\cdots(R_2-R_n)}y_2 + \cdots + \frac{(R-R_1)(R-R_2)\cdots(R-R_{n-1})}{(R_n-R_1)(R_n-R_2)\cdots(R_n-R_{n-1})}y_n \tag{12}$$

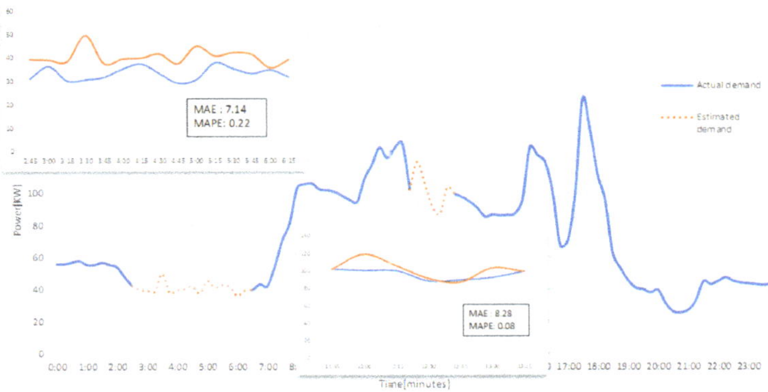

Figure 3. Estimated missing data values.

The estimated error margin for each continuous range of imputation is (MAPE: 0.22, MAE: 7.14) and (MAPE: 0.08, MAE: 8.24) for the first and second ranges, respectively. With the estimated margin errors, it is evident that the reduction method could be used to compensate for data irregularities.

3.2. Feature Variable Selection

Many predicting features affect energy consumption. For instance, date meta-data features, such as the time of day, type of day, season, and weather conditions, correlate with energy consumption. The selection of such input parameters is critical to the performance of the predictive model. Uncorrelated or associated parameters may not only lead to model overfitting but affect the performance of the predicting model [42,43]. The set of variables for this study, described in Table 1, were chosen due to their correlation with energy consumption.

Table 1. Features description.

Variable Type	Variable Name	Value
Predictors	Year	4-digit number year. E.g., 2017
	Month	2-digit number month. E.g., 01
	Day	2-digit number day. E.g., 06
	Hour	2-digits number hour. E.g., 23
	Quarter index	One digit for minutes. E.g., 1(15 min), 2(30 min)
	P1	Day of the week: 1(Mon), 2(Tues), … 7(Sun)
	P2	Day type: E.g., 1(Holidays), 2(weekdays), 3(Weekends)
	P3	The highest temperature in °C
	P4	Cloud cover: E.g. 1: sunny (cloud 0–5 mm) 2: cloudy (cloud 6–10 mm)
Respond	Demand	Energy consumption in kW

Figures 4 and 5 show the relationship between the temperature and cloud cover with the demand load. There is a positive correlation between the selected predictors and the demand load. With different load profiles for weekdays and weekends, it imperative to mention the effect that temperature and cloud cover have on the demand load. During temperate seasons, load demands are not as high as during hot seasons because higher energy consuming devices, such as air-conditioning and ventilating, will not be used as much compared with during hot seasons like summer.

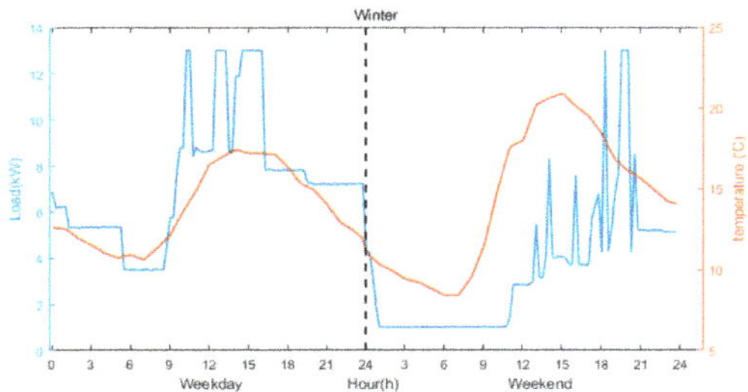

Figure 4. Energy consumption variations with temperature.

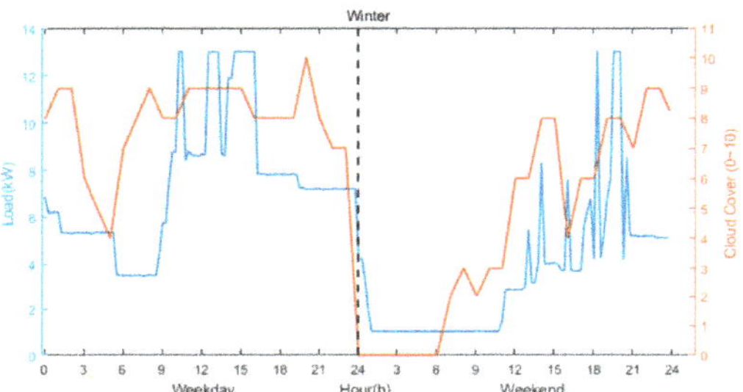

Figure 5. Energy consumption variations with cloud cover.

3.3. The Ensemble Strategy of Multiple Models

The predictive ensemble model (PEM) in this study is an adaptation of the random forest algorithm. PEM integrates series of forecasting models to formulate a forecasting model. The integration process is in two stages: optimal weight estimation and stochastic forecast result generation. The optimal weight estimation method estimates the hyperparameters of the parametric predicting models using particle swarm optimization (PSO) at each given time, as defined in Equation (13).

$$f_{e,t}(X_{n,t}) = \sum_{n=1}^{N} \omega_{n,t} \cdot f_{n,t}(X_{n,t}) \tag{13}$$

The weights ω are estimated by solving the optimization problem using the heuristic function above and the algorithm in Equation (14), where $\hat{\omega}_T$ and $\omega_{n,t}$ is the estimated optimal weight vector for the period T and weight parameter of the n-th model at the time, t, respectively.

$$\hat{\omega}_T = \underset{\omega_T}{argmin} \sum_{t \in T} L_t \left(\sum_{n=1}^{N} \omega_{n,t} \cdot f_{n,t}(X_{n,t}), y_t \right) \tag{14}$$

s.t.

$$\sum_{n=1}^{N} \omega_{n,t} = 1; \omega_{n,t} \geq 0 \; \forall \; n \in \{1, \ldots, N\} \tag{15}$$

The determination of the weights for each model per each forecast instance is initially chosen at random following the set constraints. The weights are subsequently adjusted dynamically as a result of the bias of each point forecast. With the loss function defined to estimate the error associated with each point forecast, the goal of the optimization process is to minimize the loss function, L_t, as described in Equation (16), where $X_{n,t}$ and y_t are the input parameter vector of the n-th parametric model and the actual load at time, t, respectively.

$$L_t = \sum_{n=1}^{N} (\omega_{n,t} \cdot f_{n,t}(X_{n,t}) - y_t); \forall\, t \in \{1, \cdots, T\} \tag{16}$$

The weight vector with the minimum loss function becomes the optimal weight vector. Subsequently, the optimal weight estimation is the error estimation distribution with the optimal weight vector. The stochastic forecast results are, therefore, estimated based on the error distribution and the optimal weight vector, as described in Algorithm 1.

Algorithm 1: Algorithm for Stochastic Demand Load Forecast.

Input:	Recent past actual load data Forecast period feature parameters Estimated ensemble hyperparameters Trained prediction models
Output:	Time-series stochastic load forecast
1:	Select the input parameters, $X_{n,t}$ from the set of n forecast features variables at time t.
2:	For each point-forecast model, f_n estimate the load $\hat{y}_t$, at time t with ensemble hyperparameter $\omega_{n,t}$ as shown: $\hat{y}_t = \sum_{n=1}^{N} \omega_{n,t} \cdot f_{n,t}(X_{n,t}); \forall\, t \in \{1, \cdots, T\}, \forall\, n \in \{1, \cdot\cdot, N\}$
3:	For each measured load, y_t from a set of recently measured load data, forecast the demand load, L_t and estimate the error, ε_t as follows: $\varepsilon_t = \left\| L_t - y_t \right\|; \forall\, t \in \{1, \cdots, T\}$
4:	Shift the load forecast, $\hat{y}_t$ with the error ε_t as follows: $\hat{y}_t' = \hat{y}_t + \varepsilon_t; \forall\, t \in \{1, \cdots, T\}$
5:	Fit a histogram to $\hat{y}_t'$. From the histogram, we estimate the mean, minimum and maximum value at a 95% confidence interval

---

3.4. Error Correction Model

The predictive framework proposed in this study is designed for short-term and long-term stochastic load forecasts. The framework is based on feature characteristics of historical data. However, the state characteristics of the model features may change because of uncertainties in feature variable predictions or contingencies. Thus, for each parametric model, an error correction model is defined to cater for irregularities with the forecast due to sudden changes in the feature parameters. The error correction model (ECM) is dedicated to estimating the deviations from long-term estimates to influence short-term forecasts. The ECM is defined to compensate for three error types: variance, permanent bias, and temporary bias of the prediction models. Two of the ECMs (i.e., variance and permanent bias) deal with K-means predictions, whereas the temporary bias ECM caters to ANN prediction irregularities.

3.4.1. Variance Error Correction

The traditional K-means algorithm is a vector quantization method that partitions observation into distinctive clusters. Each cluster centroid represents all members in the cluster. However, the centroid representation can lead to a high variance since each demand load profile of the cluster is represented by the same centroid, as shown in Figures 6 and 7. To compensate for such anomalies in the prediction, we defined a correction model, part of the ECM, as described in Algorithm 2, called the variance correction model. With this, the forecasted load profile was adjusted with the mean error of the recent seven-day actual demand load profile forecast. This process alters the forecasted cluster centroid with the deviations with the latest forecast to compensate for maximum variations with each cluster centroid.

Algorithm 2: Variance Error Correction Model.	
Input:	Recent past 7-days(N) actual load data
Output:	Deterministic load forecast values
1:	Select the input parameters, $X_{i,t}$ from the set of features variables at time t of load profile, $L_{i...}$
2:	Estimate the load $\hat{y}_{t,i}$, at time t of each daily load profile, L_i with K-means forecast model, f_t $$\hat{y}_{t,i} = f_t(X_{n,t}); \forall\, t \in \{1, \cdots, T\}, \forall\, n \in \{1, \cdots, N\}$$ Repeat process; $\forall\, i \in \{1, \cdots, 7\}$.
3:	For each actual load, y_t from a set of recent 7-days, measured load data, estimate the error, $\varepsilon_{t,i}$ as follows: $$\varepsilon_{t,i} = y_t - \hat{y}_{t,i}; \forall\, t \in \{1, \cdots, T\}, \forall\, i \in \{1, \cdots, N\}$$
4:	For the forecast period, t, estimate average error: $$\overline{\varepsilon_t} = \frac{1}{N} \sum_{i=1}^{7} \varepsilon_{t,i}\, i \in N, \forall\, t \in \{1, \cdots, T\}, \forall\, i \in \{1, \cdots, N\}$$
5:	Shift load forecast, $\hat{y}_t$ with mean error $\overline{\varepsilon_t}$ to form shifted load $\hat{y}'_t$ as follows: $$\hat{y}'_t = \hat{y}_t + \overline{\varepsilon_t}; \forall\, t \in \{1, \cdots, T\}$$

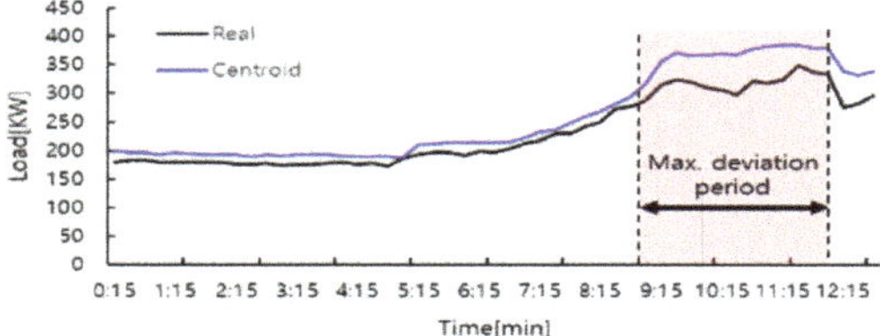

Figure 6. 22 June 2017 demand load profile.

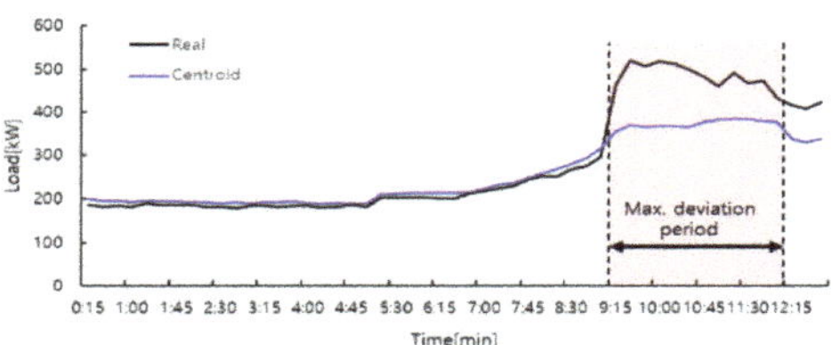

Figure 7. 30 June 2017 demand load profile.

3.4.2. Permanent Bias Error Correction

The demand load profile varies with time, even with a similar set of predictors, as shown in Table 2. The prediction gap can be significant, as depicted in Figure 8. These changes are mainly due to the addition of power equipment or changes in the number of users. As mentioned earlier, the information on the feature state characteristics is not accounted for by the K-means model.

Table 2. A set of similar load predictors.

Date	P1	P2	P3	P4
23 June 2017	5	4	34.8	1
19 July 2018	4	4	34.7	1

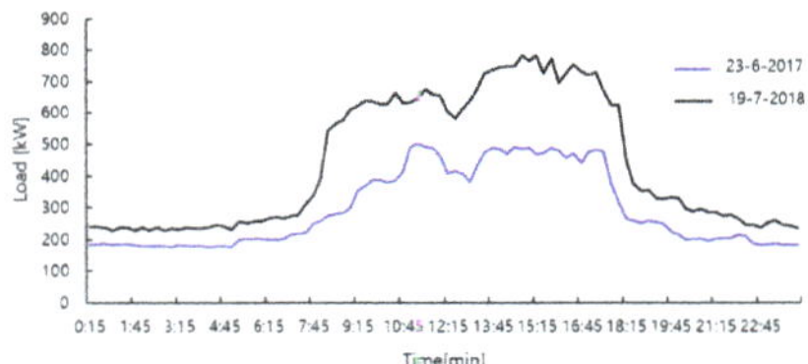

Figure 8. Comparing two demand load profiles.

A permanent bias error correction model is, therefore, devised to compensate for the variations in the demand load profile. With an intermittent short-term (i.e., 15 min) forecast interval, the correction model uses an estimated error rate of the actual data of the previous seven days to retard the forecast deviation, as shown in Figure 9. The algorithm for this process is described in Algorithm 3. The variance error correction precedes the permanent bias error correction.

Algorithm 3: Permanent Bias Error Correction Model

Input:	Recent past 7-days(N) actual load data
Output:	Deterministic load forecast values
1:	Select the input parameters, $X_{i,t}$ from the set of features variables at time t of load profile, L_i.
	Estimate the load $\hat{y}_{t,i}$, at time t of each daily load profile, L_i with K-means forecast model, f_t
2:	$$\hat{y}_{t,i} = f_t(X_{n,t}); \forall\, t \in \{1, \cdots, T\}, \forall\, n \in \{1, \cdots, N\}$$
	Repeat process; $\forall\, i \in \{1, \cdots, 7\}$.
	For each actual load, y_t from a set of recent 7-days, measured load data, estimate the error, $\varepsilon_{t,i}$
3:	as follows:
	$$\varepsilon_{t,i} = y_t - \hat{y}_{t,i}; \forall\, t \in \{1, \cdots, T\}, \forall\, i \in \{1, \cdots, N\}$$
	For the forecast period, t, estimate average error:
4:	$$\overline{\varepsilon_t} = \frac{1}{N} \sum_{i=1}^{7} \varepsilon_{t,i}$$
	$$i \in N, \forall\, t \in \{1, \cdots, T\}, \forall\, i \in \{1, \cdots, N\}$$
5:	For all $t \in T$, if $\overline{\varepsilon_t} \not< 0$ or $\overline{\varepsilon_t} \not> 0$, then estimate error rate, ε_r :
	$$\varepsilon_r = \frac{\overline{\varepsilon_t}}{y_t}; t \in \{1, \cdots, T\}, \forall\, r \in \{1, \cdots, N\}$$
6:	Shift load forecast, $\hat{y}_t$ with error rate ε_r to form shifted load $\hat{y}'_t$ as follows:
	$$\hat{y}'_t = \frac{\hat{y}_t}{1 - \varepsilon_r}; t \in \{1, \cdots, T\}, \forall\, r \in \{1, \cdots, N\}$$

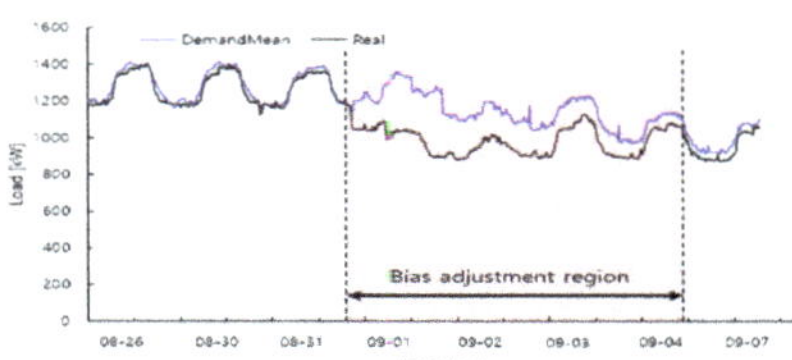

Figure 9. Permanent bias error correction effect.

3.4.3. Temporary Bias Error Correction

Special event bias may occur with persistent multiple error events even after the application of the error correction models mentioned above, as shown in Figure 10. Special event bias is defined as a sudden change in the amount of power over a quarter of the maximum load happening in a two-or-more-hour period. The evidence of such occurrences is significant in most demand load data. A classic example is the Korean research institute (KEPRI) building demand load data used in our case study analysis. In the dataset, the total number of such occurrences was 91 over a 553-day period.

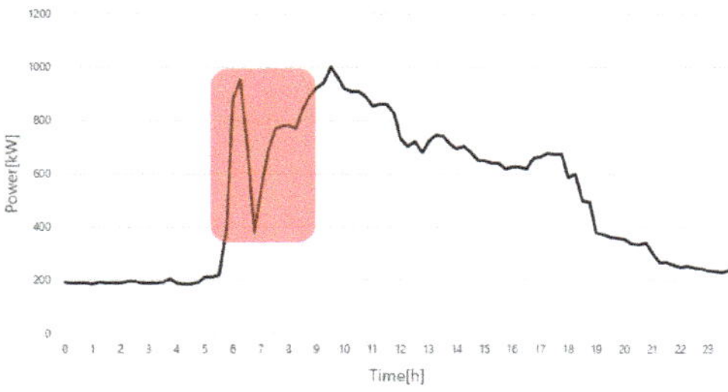

Figure 10. Special event bias.

Although it is sudden, the occurrence of special event bias is significant in a given period. From Figure 11, the duration of the sudden bias is logarithmic in time. The highest magnitude of the bias lasts for the first few hours and monotonically decreases with time to the nominal value. In this study, we compensated for these sudden changes in the demand load forecast with a temporal bias error correction model. The procedure for the error compensation is similar to the permanent bias error correction, except for the required input data and error correction formula. The occurrence distribution and occurrence duration distribution are shown in Figures 12 and 13, respectively.

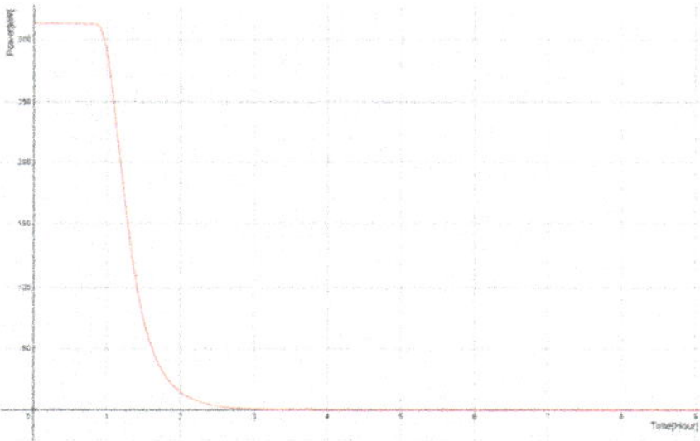

Figure 11. Special event occurrence.

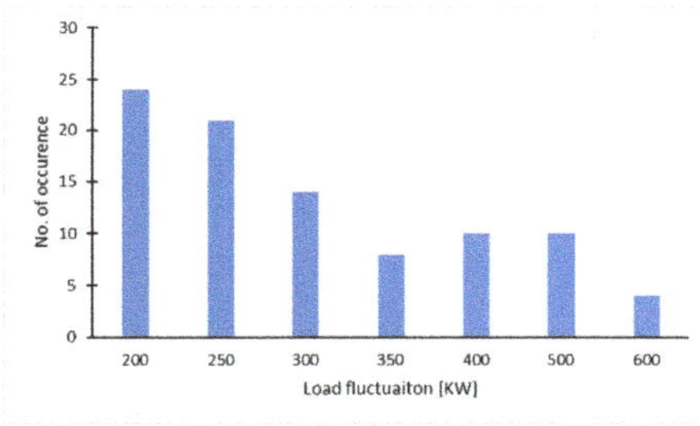

Figure 12. Special event load fluctuation distribution.

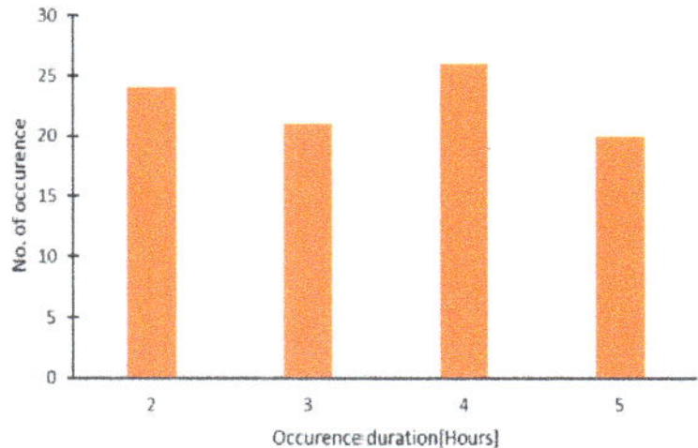

Figure 13. Special duration distribution.

Actual data for the previous 3 h (T) was utilized to estimate the error for compensation. The formulation of the correction model, as defined in Equation (17), is to correct the average quantum of the special event, learned for one hour until the power remains zero. An initial correction value, α, is an average of the event values that occurred in the past. With the KEPRI demand dataset, α was estimated as 312.6 KW. Considering that temporary bias error correction is a model for accident events, the average elapsed period, T, was set to be 3.5 h, which is the average time required to identify an event.

$$\hat{y}_t = \alpha(1 - e^{-\frac{T-t}{t^5}})$$
(17)

4. Case Study and Scenario Analysis

In this section, we show the performance of the proposed framework. In the first part, we test the algorithm on the Korean Electric Power Company (KEPCO) load consumption data from the Korean research institute (KEPRI) building and substation. The load data and weather information used were acquired from iSmart [44] and K-weather [45], respectively. We compare our forecast results with parametric model forecasts. In the second part, we test the algorithm on a set of generalized testbed data of different load sizes, shapes, and characteristics. Smart energy meters and IoT sensors were used to gather load data and feature data for the prediction analysis. Prediction analysis for four seasons—spring, summer, fall, and winter—was conducted in each case study. The optimal values estimated were used to predict randomly selected days for each season. In this study, the average model training computational time and predictive time are specified in Table 3. All the data processing and modeling tasks were implemented using MATLAB software (2019a, MathsWorks, Natick, MA, USA) on a 64-bit Intel i7 (4 CPUs and 16 GB RAM) with Windows operating system. The proposed model is relatively computationally non-intensive for both online and offline load forecasts. However, the training computation time could be reduced with parallel processing.

Table 3. The computation time of processes.

Process	Model Training [s]	Forecast [s]
K-means	50.86	1.44
ANN	61.70	20.58
Ensemble	592.75	0.002

4.1. Case I: Performance of the Proposed Model on Korea Power Company Buildings Dataset

Our predictive analysis, first and foremost, begins with the electrical load dataset from KEPRI building and KEPCO dong-Daegu substation building for which historical data is available from 2015 to date, as shown in Figures 14 and 15 respectively. Based on the available accumulated historical demand load and weather information data, a prediction model was developed. We estimated the prediction accuracy by predicting demand loads of 2019, considering both the MAPE and RMSE indices.

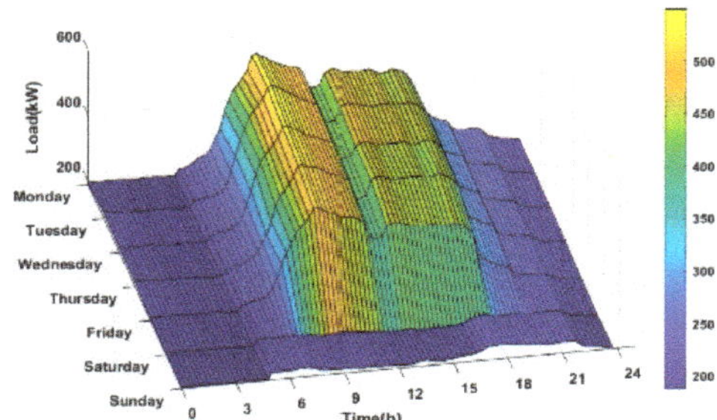

Figure 14. Korean research institute (KEPRI) demand load profile.

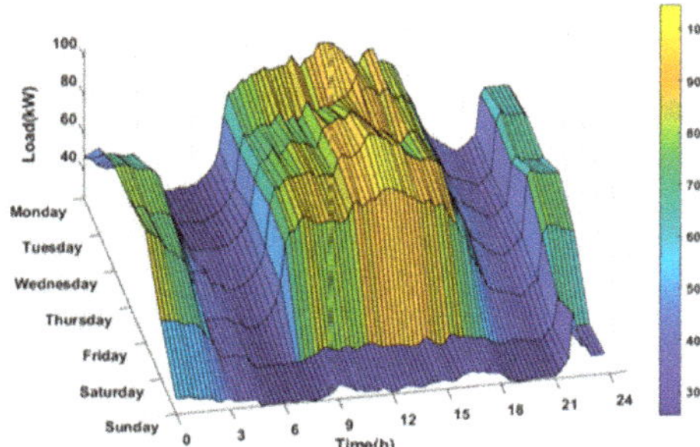

Figure 15. Korean Electric Power Company (KEPCO) substation demand load profile.

The first stage of the analysis was to analyze the scenarios of two datasets (i.e., KEPRI and KEPCO) of different feature characteristics. These datasets contain demand load and feature data at a 15-minute sample resolution. The datasets cover the period of five years from September 2015. A four-year dataset was used to train the predictive model, whereas a one-year dataset was used for testing and cross-validation. The hyperparameters were tuned to compensate for error correction using the seven days prior to the starting date. Specific days among the four seasons were selected for load prediction. Figures 16 and 17 show the stochastic forecast results of the ensemble prediction for the selected days in each season of the two datasets above.

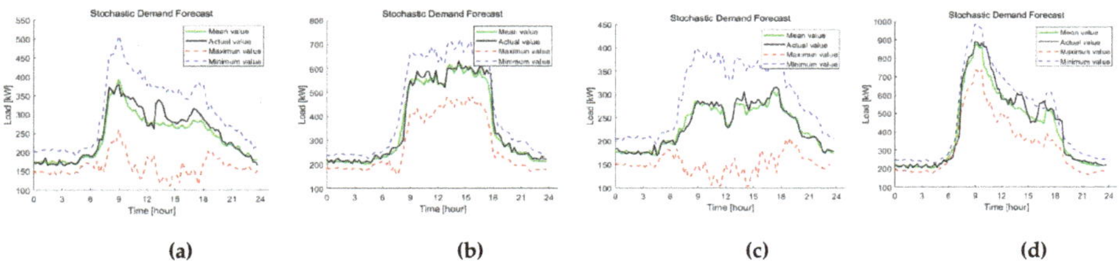

Figure 16. KEPRI stochastic demand forecast. (**a**) Spring, (**b**) Summer, (**c**) Fall, (**d**) Winter.

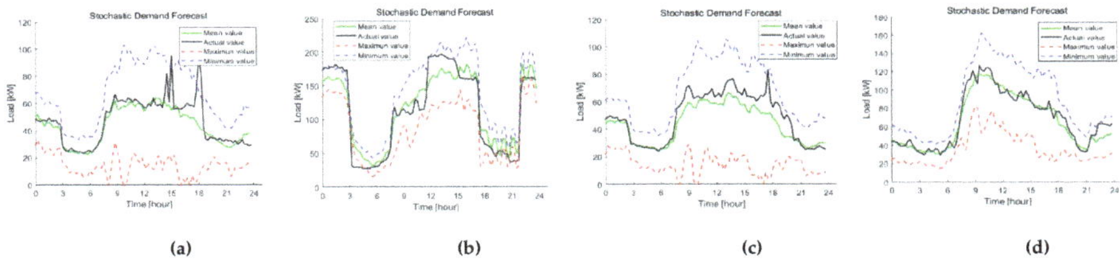

Figure 17. KEPCO substation stochastic demand forecast. (**a**) Spring, (**b**) Summer, (**c**) Fall, (**d**) Winter.

From the result, it is imperative to know that all seasons take the shape and form of the actual measured load profile. In most cases, the actual measured values lie within the minimum and maximum confidence interval. Before reporting the performance of the ensemble model obtained by combining multiple individual models, we first analyzed the performance of the parametric models mentioned in Section 3. With the same dataset for both training and validation, we estimated the demand load forecast with ANN and K-means separately without the ensemble effect.

To yield credible results, we trained each model multiple times and averaged the losses. From the prediction error results, as shown in Figure 18 and Table 4, it is observed in the figures that the ensemble model can improve the performance of the model, with the error correction model implementation. For simplicity, the ensemble model consisted of only two parametric models, but the framework is scalable for multiple models.

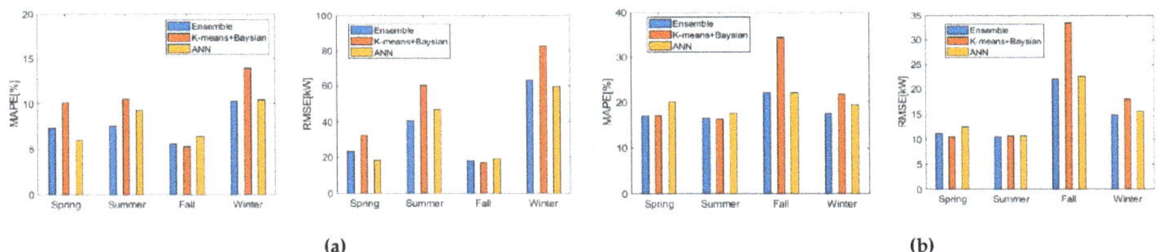

Figure 18. Performance evaluation of prediction models. (**a**) KEPRI data set, (**b**) KEPCO dataset.

Table 4. Prediction accuracy of forecasting dataset.

Dataset			Ensemble	K-Means with Bayesian	ANN
KEPCO dataset	MAPE	Spring	17.0928	17.06893	20.04814
		Summer	16.45506	16.33028	17.51668
		Fall	22.02287	34.34863	21.94459
		Winter	17.57218	21.73072	19.42727
	RMSE	Spring	11.19016	10.59841	12.52269
		Summer	10.55273	10.79871	10.73513
		Fall	22.00363	33.5403	22.5707
		Winter	14.8416	17.97763	15.57768
KEPRI dataset	MAPE	Spring	7.3435	10.1011	6.013
		Summer	7.52238	10.4581	9.22479
		Fall	5.52856	5.27346	6.41714
		Winter	10.2523	13.90402	10.38853
	RMSE	Spring	23.41024	32.69604	18.52118
		Summer	40.75116	60.16406	46.72084
		Fall	18.07842	17.08345	19.13558
		Winter	63.23432	82.75472	59.46993

4.2. Case II: Performance of the Proposed Model on Testbed Dataset

The second scenario of the case studies was to examine the adaptability of the proposed model. To this end, we use the hyper-parameters of the ensemble model tuned with the testbed dataset to train load forecasting models for different load form forecasts. Here, the load profile of the dataset does not follow any regular pattern, unlike in Case I. The testbed platform gathers information from a living laboratory, which consists of research laboratories, teaching classrooms, and faculty offices. The activities in these places are not routine; hence, the load profile does not follow a regular pattern, as shown in Figure 19.

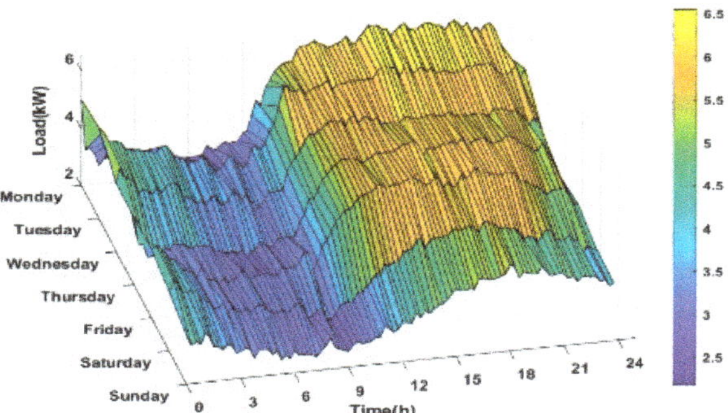

Figure 19. Testbed demand load profile.

For the proposed ensemble model, the training period was from June 2015 to December 2018. Similar to Case I, neural network and K-means parametric models, together with the proposed ensemble model, were used to train demand data independently from 2015 to May 2019, while the data from June 2019 to December 2019 was used for the forecast. Figures 20 and 21 show the results of the stochastic forecast and prediction accuracy, respectively. It is seen in the figures that the proposed ensemble model has the lowest overall MAPE and RMSE among the predictive models.

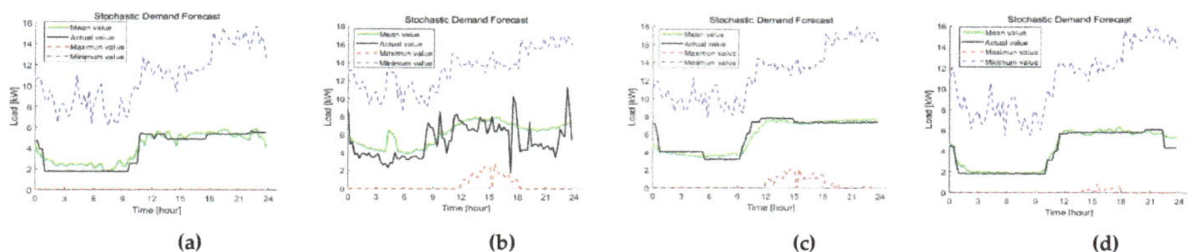

Figure 20. Testbed stochastic demand load forecast. (**a**) Spring, (**b**) Summer, (**c**) Fall, (**d**) Winter.

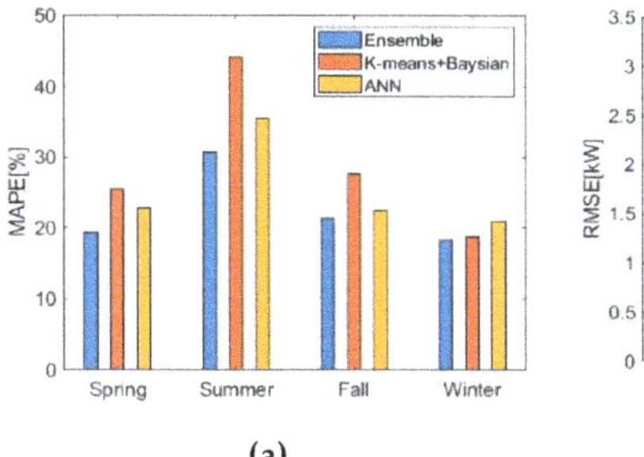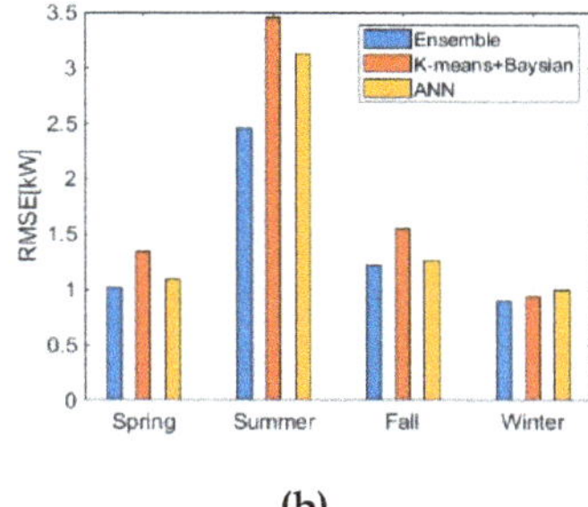

(a)   (b)

Figure 21. Testbed dataset performance evaluation of prediction models. (**a**) Performance estimation with MAPE, (**b**) Performance estimation with RMSE.

For all three prediction models, we repeated the trials for randomly selected days and estimated the average MAPE and RMSE. As shown in Figure 21 and Table 5, the ensemble strategy dramatically reduces the losses of MAPE and RMSE, especially for different load profiles with no regular pattern. It is, therefore, reasonable to conclude that the proposed model is robust even for demand load with many irregularities in load patterns. It is also observed that the ensemble strategy can reduce the deviation of multiple trials. This also indicates the higher adaptive capability of the proposed model with the ensemble strategy.

Table 5. Prediction accuracy of forecasting testbed dataset.

Performance Index		Ensemble	K-Means with Bayesian	ANN
MAPE	Spring	19.27277	25.50984	22.78469
	Summer	30.74737	44.09545	35.52047
	Fall	21.35013	27.61316	22.40761
	Winter	18.30707	18.67867	20.87263
RMSE	Spring	1.01556	1.34224	1.085257
	Summer	2.455532	3.452222	3.130337
	Fall	1.21539	1.541551	1.264588
	Winter	0.892299	0.938574	0.999332

5. Conclusions

This paper proposes an ensemble prediction framework for stochastic demand load forecasting to take advantage of the diversities in forecasting models. The ensemble problem is formulated as a two-stage random forest problem with a series of homogenous prediction models. A heuristic trained model combiner, together with an error correction model, enable the proposed model to have high accuracy as well as satisfactory adaptive capabilities. KEPRI building, KEPCO substation building, and testbed datasets were used to verify the effectiveness of the proposed model with different case scenarios. The comparisons with existing parametric models show that the proposed model is superior in both forecasting accuracy and robustness in load profile variation. The results from the case study analysis show that the proposed ensemble method effectively improves the forecasting performance when compared with individual models. For simplicity, two parametric models(i.e., K-means with Bayesian and ANN models) were adapted in this analysis. However, in future work, other deep learning models, such as a recurrent neural network (RNN) and long short-term memory (LSTM) neural network, can be seamlessly incorporated into the model to enhance its performance.

Author Contributions: Conceptualization, S.H. and K.A.A.; Methodology, K.A.A.; Software, K.A.A.; Validation, S.P., and G.K. Formal Analysis, K.A.A., G.K., S.P. and H.J.; Investigation, G.K., H.J. and S.P.; Resources, S.H.; Data Curation, K.A.A. and G.K. and S.P.; Writing-Original Draft Preparation, K.A.A.; Writing-Review & Editing, K.A.A.; Visualization, K.A.A. and G.K.; Supervision, S.H.; Project Administration, S.H.; Funding Acquisition, S.H. All authors have read and agreed to the published version of the manuscript.

Funding: This research was funded by Korea Institute of Energy Technology Evaluation and Planning(KETEP) and the Ministry of Trade, Industry and Energy(MOTIE) grant number [20182010600390].

Conflicts of Interest: The funders had no role in the design of the study; in the collection, analyses, or interpretation of data; in the writing of the manuscript, and in the decision to publish the results.

References

1. Spiliotis, K.; Ramos Gutierrez, A.I.; Belmans, R. Demand flexibility versus physical network expansions in distribution grids. *Appl. Energy* **2016**, *182*, 613–624. [CrossRef]
2. Fitariffs. Feed-in Tariffs. Available online: https://www.fitariffs.co.uk/fits/ (accessed on 14 January 2019).
3. Nosratabadi, S.M.; Hooshmand, R.-A.; Gholipour, E. A comprehensive review on microgrid and virtual power plant concepts employed for distributed energy resources scheduling in power systems. *Renew. Sustain. Energy Rev.* **2017**, *67*, 341–363. [CrossRef]
4. Fan, C.; Xiao, F.; Wang, S. Development of prediction models for next-day building energy consumption and peak power demand using data mining techniques. *Appl. Energy* **2014**, *127*, 1–10. [CrossRef]
5. Xenos, D.P.; Mohd Noor, I.; Matloubi, M.; Cicciotti, M.; Haugen, T.; Thornhill, N.F. Demand-side management and optimal operation of industrial electricity consumers: An example of an energy-intensive chemical plant. *Appl. Energy* **2016**, *182*, 418–433. [CrossRef]
6. Sepehr, M.; Eghtedaei, R.; Toolabimoghadam, A.; Noorollahi, Y.; Mohammadi, M. Modeling the electrical energy consumption profile for residential buildings in Iran. *Sustain. Cities Soc.* **2018**, *41*, 481–489. [CrossRef]
7. Tucci, M.; Crisostomi, E.; Giunta, G.; Raugi, M. A Multi-Objective Method for Short-Term Load Forecasting in European Countries. *IEEE Trans. Power Syst.* **2016**, *31*, 3537–3547. [CrossRef]
8. Wang, C.-h.; Grozev, G.; Seo, S. Decomposition and statistical analysis for regional electricity demand forecasting. *Energy* **2012**, *41*, 313–325. [CrossRef]
9. Calderón, C.; James, P.; Urquizo, J.; McLoughlin, A. A GIS domestic building framework to estimate energy end-use demand in UK sub-city areas. *Energy Build.* **2015**, *96*, 236–250. [CrossRef]
10. Raza, M.Q.; Khosravi, A. A review on artificial intelligence based load demand forecasting techniques for smart grid and buildings. *Renew. Sustain. Energy Rev.* **2015**, *50*, 1352–1372. [CrossRef]
11. Wang, Q.; Zhou, B.; Li, Z.; Ren, J. Forecasting of short-term load based on fuzzy clustering and improved BP algorithm. In Proceedings of the 2011 International Conference on Electrical and Control Engineering, Yichang, China, 16–18 September 2011; pp. 4519–4522.
12. Hernandez, L.; Baladron, C.; Aguiar, J.M.; Carro, B.; Sanchez-Esguevillas, A.; Lloret, J.; Chinarro, D.; Gomez-Sanz, J.J.; Cook, D. A multi-agent system architecture for smart grid management and forecasting of energy demand in virtual power plants. *IEEE Commun. Mag.* **2013**, *51*, 106–113. [CrossRef]
13. Hippert, H.S.; Pedreira, C.E.; Souza, R.C. Neural networks for short-term load forecasting: A review and evaluation. *IEEE Trans. Power Syst.* **2001**, *16*, 44–55. [CrossRef]
14. Kaytez, F.; Taplamacioglu, M.C.; Cam, E.; Hardalac, F. Forecasting electricity consumption: A comparison of regression analysis, neural networks and least squares support vector machines. *Int. J. Electr. Power Energy Syst.* **2015**, *67*, 431–438. [CrossRef]
15. Jain, R.K.; Smith, K.M.; Culligan, P.J.; Taylor, J.E. Forecasting energy consumption of multi-family residential buildings using support vector regression: Investigating the impact of temporal and spatial monitoring granularity on performance accuracy. *Appl. Energy* **2014**, *123*, 168–178. [CrossRef]
16. Kuo, R.J.; Li, P.S. Taiwanese export trade forecasting using firefly algorithm based K-means algorithm and SVR with wavelet transform. *Comput. Ind. Eng.* **2016**, *99*, 153–161. [CrossRef]
17. Zhang, F.; Deb, C.; Lee, S.E.; Yang, J.; Shah, K.W. Time series forecasting for building energy consumption using weighted Support Vector Regression with differential evolution optimization technique. *Energy Build.* **2016**, *126*, 94–103. [CrossRef]
18. Massana, J.; Pous, C.; Burgas, L.; Melendez, J.; Colomer, J. Short-term load forecasting for non-residential buildings contrasting artificial occupancy attributes. *Energy Build.* **2016**, *130*, 519–531. [CrossRef]

19. Sandels, C.; Widén, J.; Nordström, L.; Andersson, E. Day-ahead predictions of electricity consumption in a Swedish office building from weather, occupancy, and temporal data. *Energy Build.* **2015**, *108*, 279–290. [CrossRef]

20. Wang, X.; Lee, W.; Huang, H.; Szabados, R.L.; Wang, D.Y.; Olinda, P.V. Factors that Impact the Accuracy of Clustering-Based Load Forecasting. *IEEE Trans. Ind. Appl.* **2016**, *52*, 3625–3630. [CrossRef]

21. Deihimi, A.; Orang, O.; Showkati, H. Short-term electric load and temperature forecasting using wavelet echo state networks with neural reconstruction. *Energy* **2013**, *57*, 382–401. [CrossRef]

22. Mena, R.; Rodrguez, F.; Castilla, M.; Arahal, M. A prediction model based on neural networks for the energy consumption of a bioclimatic building. *Energy Build.* **2014**, *82*, 142–155. [CrossRef]

23. Wang, Y.; Chen, Q.; Sun, M.; Kang, C.; Xia, Q. An Ensemble Forecasting Method for the Aggregated Load With Subprofiles. *IEEE Trans. Smart Grid* **2018**, *9*, 3906–3908. [CrossRef]

24. Wang, Y.; Zhang, N.; Tan, Y.; Hong, T.; Kirschen, D.S.; Kang, C. Combining Probabilistic Load Forecasts. *IEEE Trans. Smart Grid* **2019**, *10*, 3664–3674. [CrossRef]

25. Chen, Y.; Tan, H.; Berardi, U. Day-ahead prediction of hourly electric demand in non-stationary operated commercial buildings: A clustering-based hybrid approach. *Energy Build.* **2017**, *148*, 228–237. [CrossRef]

26. Fumo, N.; Mago, P.; Luck, R. Methodology to estimate building energy consumption using EnergyPlus Benchmark Models. *Energy Build.* **2010**, *42*, 2331–2337. [CrossRef]

27. Ji, Y.; Xu, P.; Ye, Y. HVAC terminal hourly end-use disaggregation in commercial buildings with Fourier series model. *Energy Build.* **2015**, *97*, 33–46. [CrossRef]

28. Bracale, A.; Caramia, P.; Carpinelli, G.; Fazio, A.R.D.; Varilone, P. A Bayesian-Based Approach for a Short-Term Steady-State Forecast of a Smart Grid. *IEEE Trans. Smart Grid* **2013**, *4*, 1760–1771. [CrossRef]

29. Collotta, M.; Pau, G. An Innovative Approach for Forecasting of Energy Requirements to Improve a Smart Home Management System Based on BLE. *IEEE Trans. Green Commun. Netw.* **2017**, *1*, 112–120. [CrossRef]

30. Deng, Z.; Wang, B.; Xu, Y.; Xu, T.; Liu, C.; Zhu, Z. Multi-Scale Convolutional Neural Network With Time-Cognition for Multi-Step Short-Term Load Forecasting. *IEEE Access* **2019**, *7*, 88058–88071. [CrossRef]

31. Ding, N.; Benoit, C.; Foggia, G.; Bésanger, Y.; Wurtz, F. Neural Network-Based Model Design for Short-Term Load Forecast in Distribution Systems. *IEEE Trans. Power Syst.* **2016**, *31*, 72–81. [CrossRef]

32. Han, L.; Peng, Y.; Li, Y.; Yong, B.; Zhou, Q.; Shu, L. Enhanced Deep Networks for Short-Term and Medium-Term Load Forecasting. *IEEE Access* **2019**, *7*, 4045–4055. [CrossRef]

33. Li, R.; Li, F.; Smith, N.D. Multi-Resolution Load Profile Clustering for Smart Metering Data. *IEEE Trans. Power Syst.* **2016**, *31*, 4473–4482. [CrossRef]

34. Motepe, S.; Hasan, A.N.; Stopforth, R. Improving Load Forecasting Process for a Power Distribution Network Using Hybrid AI and Deep Learning Algorithms. *IEEE Access* **2019**, *7*, 82584–82598. [CrossRef]

35. Ouyang, T.; He, Y.; Li, H.; Sun, Z.; Baek, S. Modeling and Forecasting Short-Term Power Load With Copula Model and Deep Belief Network. *IEEE Trans. Emerg. Top. Comput. Intell.* **2019**, *3*, 127–136. [CrossRef]

36. Tang, X.; Dai, Y.; Wang, T.; Chen, Y. Short-term power load forecasting based on multi-layer bidirectional recurrent neural network. *IET Gener. Transm. Distrib.* **2019**, *13*, 3847–3854. [CrossRef]

37. Teixeira, J.; Macedo, S.; Gonçalves, S.; Soares, A.; Inoue, M.; Cañete, P. Hybrid model approach for forecasting electricity demand. *CIRED Open Access Proc. J.* **2017**, *2017*, 2316–2319. [CrossRef]

38. Xu, T.; Chiang, H.; Liu, G.; Tan, C. Hierarchical K-means Method for Clustering Large-Scale Advanced Metering Infrastructure Data. *IEEE Trans. Power Deliv.* **2017**, *32*, 609–616. [CrossRef]

39. Zhang, W.; Quan, H.; Srinivasan, D. An Improved Quantile Regression Neural Network for Probabilistic Load Forecasting. *IEEE Trans. Smart Grid* **2019**, *10*, 4425–4434. [CrossRef]

40. Khan, I.; Capozzoli, A.; Corgnati, S.P.; Cerquitelli, T. Fault Detection Analysis of Building Energy Consumption Using Data Mining Techniques. *Energy Procedia* **2013**, *42*, 557–566. [CrossRef]

41. Carsey, V.J.; Wagner, C.G.; Walters, E.E.; Rosner, B.A. Resistant and test based outlier rejection: Effects on Gaussian one- and two-sample inference. *Technometrics* **1997**, *39*, 320–330. [CrossRef]

42. Jovanovic, R.Z.; Sretenovic, A.A.; Zivkovic, B.D. Ensemble of various neural networks for prediction of heating energy consumption. *Energy Build.* **2015**, *94*, 189–199. [CrossRef]

43. Zhang, Y.N.; O'Neill, Z.; Dong, B.; Augenbroe, G. Comparisons of inverse modeling approaches for predicting building energy performance. *Build. Environ.* **2015**, *86*, 177–190. [CrossRef]

44. KEPCO. iSmart-Smart Power Management. Available online: https://pccs.kepco.co.kr/iSmart/ (accessed on 4 August 2017).
45. Kweather. Kweather-Total Weather Service Provider. Available online: http://www.kweather.co.kr/main/main.html (accessed on 6 March 2016).

Article

Where Will You Park? Predicting Vehicle Locations for Vehicle-to-Grid

Rob Shipman *, Julie Waldron, Sophie Naylor, James Pinchin, Lucelia Rodrigues and Mark Gillott

Department of Architecture and Built Environment, Faculty of Engineering, The University of Nottingham, University Park, Nottingham NG7 2RD, UK; julie.waldron@nottingham.ac.uk (J.W.); sophie.naylor@nottingham.ac.uk (S.N.); james.pinchin@nottingham.ac.uk (J.P.); lucelia.rodrigues@nottingham.ac.uk (L.R.); mark.gillott@nottingham.ac.uk (M.G.)
* Correspondence: rob.shipman@nottingham.ac.uk; Tel.: +44-1157486721

Received: 27 March 2020; Accepted: 12 April 2020; Published: 14 April 2020

Abstract: Vehicle-to-grid services draw power or curtail demand from electric vehicles when they are connected to a compatible charging station. In this paper, we investigated automated machine learning for predicting when vehicles are likely to make such a connection. Using historical data collected from a vehicle tracking service, we assessed the technique's ability to learn and predict when a fleet of 48 vehicles was parked close to charging stations and compared this with two moving average techniques. We found the ability of all three approaches to predict when individual vehicles could potentially connect to charging stations to be comparable, resulting in the same set of 30 vehicles identified as good candidates to participate in a vehicle-to-grid service. We concluded that this was due to the relatively small feature set and that machine learning techniques were likely to outperform averaging techniques for more complex feature sets. We also explored the ability of the approaches to predict total vehicle availability and found that automated machine learning achieved the best performance with an accuracy of 91.4%. Such technology would be of value to vehicle-to-grid aggregation services.

Keywords: vehicle-to-grid; V2G; vehicle location prediction; automated machine learning; machine learning

---

1. Introduction

A key function of an electricity grid operator is to balance supply and demand to ensure that the power produced always matches the power required. In the UK, for example, this is achieved by the Balancing Mechanism of the National Grid, which calculates deviations in supply and demand every half-hour. To address any imbalances, the operator will accept offers to increase or curtail demand and/or generation in near real-time. Electricity can also be traded ahead of time; in the day-ahead market, for example, generators and suppliers agree contracts for the delivery of energy typically during hour periods on the following day [1]. Vehicle-to-grid (V2G) is a technology that allows electric vehicles to contribute to such flexibility services by discharging or curtailing demand when required [2,3]. This capability has the potential to help manage the additional load on the grid resulting from the influx of electric vehicles, to help manage supply fluctuations inherent to renewable energy sources and to contribute to ambitious sustainability targets introduced by many cities around the world, including Nottingham in the UK [4].

While the integration of static energy storage within virtual power plants is relatively well developed [5], significant additional challenges result where the storage is mobile in the form of electric vehicles (EVs). For example, charging and discharging must be scheduled and aligned with vehicle availability, and use of the battery must respect the primary use of the vehicle as a form of transport. Commercial organisations have been established to offer such capability [6] attracted by the significant opportunities offering flexibility services to the electricity grid [7]. Energy companies, such as Octopus Energy [8] and Ovo Energy [9] in the UK, are now also rolling-out services based on V2G.

Participation in market opportunities is, however, reliant on the availability of enough vehicles at the time of the market event. As the total population of participating vehicles grows, it becomes more likely that enough vehicles would be available, given that many are typically parked over 95% of the time [10]. However, as trading decisions are typically made in advance, finer-grained predictions of available capacity become necessary, and support participation in larger and more numerous market events as a smaller buffer of vehicles is required to account for uncertainty. Such predictions also enable the use of the technology for scenarios with an inherently smaller vehicle population, such as individual communities or local vehicle-to-building applications [11].

A prediction of available capacity is critically dependent on many factors, including battery capacity and state-of-charge; however, fundamental to this prediction is the actual availability of the vehicle, i.e., it must be parked close enough to an available charging station to be plugged in. This, therefore, requires predicting the stationary location of vehicles—a problem that has been explored previously in the literature. Markov models, for example, have been used to model driving patterns using a single vehicle's data [12] and to model a vehicle's state using survey data [13]. The related problems of travel time prediction [14] and parking space prediction [15] have also received considerable attention. However, to enable V2G services, there remains a need for techniques to predict when vehicles are parked close enough to charging stations and hence potentially available to a V2G aggregation service. These techniques must also be validated using real data from a substantial number of vehicles.

In this paper, we addressed this need by using a historical dataset from a fleet of vehicles to train and analysed several different predictive models. We made the following specific contributions; firstly, we demonstrated the ability of the models to predict when vehicles are parked close to V2G charging stations with high accuracy, which is necessary to underpin the assessment of the capacity available to a V2G aggregation service during future trading windows; secondly, we demonstrated a method of analysing a dataset retrieved from a vehicle tracking service to support the identification of vehicles that are strong candidates for use in a V2G service; thirdly, we demonstrated that simple prediction strategies, such as moving averages, could yield comparable performance to more complex machine learning techniques, which is of value to help bootstrap V2G services when large training datasets are not initially available.

The remainder of the paper is structured as follows; in Section 2, we described the dataset used to train the models and detailed the three approaches investigated; in Section 3, we compared, analysed and discussed the performance of the approaches in predicting the availability of individual vehicles and total available vehicles; Section 4 presents our conclusions.

2. Materials and Methods

In this work, we used 42 weeks of historical data from a fleet of 48 vehicles belonging to the University of Nottingham that was collected using the Trakm8 telematics service [16] deployed in those vehicles. We investigated the use of automated machine learning [17] (AutoML) that has the potential to broaden the use of machine learning within the energy domain by automating the time-consuming workflow and allowing the rapid exploration of a range of industry-standard algorithms. This technique was compared with two averaging techniques: a simple cumulative moving average (CMA) and an exponential moving average (EMA) that weights recent data more strongly. We assessed the ability of the three approaches to predict the availability of individual vehicles and the total available vehicles in future half-hour periods, i.e., potential trading windows.

2.1. Dataset Processing

The University of Nottingham operates a fleet of 121 vehicles across 4 UK campuses, which provide a wide variety of roles, including catering services, estates management and security. A total of 48 of these vehicles from 6 different departments were actively tracked using the Trakm8 service, which provided detailed information on vehicle condition, driving patterns and individual journey details. The latter included the time and GPS location at the start and end of the journey from which latitude and longitude could be derived, as shown in Table 1. Analysis of this data thus allowed a dataset to be constructed of when, where and for how long each vehicle was stationary.

Table 1. Example data received for each vehicle journey.

Name	Description	Example
vid	Unique identifier for this vehicle	12
start_lat	Latitude at the start of the journey	52.95282
start_lng	Longitude at the start of the journey	−1.18652
start_time	Timestamp at the start of the journey	2019-11-21T13:53:10+00:00
end_lat	Latitude at the end of the journey	52.94025
end_lng	Longitude at the end of the journey	−1.192132
end_time	Timestamp at the end of the journey	2019-11-21T14:00:16+00:00

At the time of the study, the fleet was not equipped with V2G technology, and the compatible charge points were not available. However, the best potential locations for V2G charge points were determined through a combination of (i) interviews with fleet managers to understand the patterns of use of the vehicles and overnight parking location of the different fleets, (ii) analysis of Trakm8 data to identify typical parking locations of the tracked vehicles, (iii) assessing infrastructure feasibility to install V2G chargers (e.g., energy supply availability to connect 3-phase V2G chargers) [18]. This analysis resulted in the identification of 6 proposed locations spread across 3 campuses in the city of Nottingham, UK.

Cross-referencing parked locations with each of these charge point locations allowed the number of vehicles to be determined that could potentially be available if the necessary hardware was in place. This was achieved by calculating the great-circle distance using the haversine formula, as shown in Equation (1), where r is the radius of the earth (6371 km), and $dist_i$ is the distance in km between the location of a parked vehicle v (end_lat_v and end_lng_v) and charger location i (lat_i and lng_i).

$$dist_i = 2 * r * \arcsin\left(\sqrt{\sin^2\left(\frac{lat_i - end\_lat_v}{2}\right) + \cos(end\_lat_v) * \cos(lat_i) * \sin^2\left(\frac{lng_i - end\_lng_v}{2}\right)} \right) \quad (1)$$

When the shortest distance to a charge point was below 100 m, the vehicle was considered to be parked within a suitable radius and hence potentially available to a V2G aggregation service, i.e., $a_v = 1$, as shown in Equation (2). This radius was chosen to account for inevitable variance in GPS locations and to be close enough to require only minor changes in behaviour to park close enough to a charging station to be plugged in, e.g., choosing a different parking place within the same car park.

$$(\min\{dist_i\}_{i=1}^6 < 0.1 \rightarrow a_v = 1) \wedge (\min\{dist_i\}_{i=1}^6 \geq 0.1 \rightarrow a_v = 0) \quad (2)$$

Forty-two weeks of data were collected, and each of the 294 days, d, represented in the dataset was divided into 48 contiguous half-hour periods; hh_i^d, $1 \leq i \leq 48$, $1 \leq d \leq 294$. The dataset was then processed to determine vehicle availability as follows:

For each pair of consecutive journeys, J_n^v and J_{n+1}^v, in the dataset for each vehicle, v:

- The stationary period, p, was calculated as the set of full minutes between the end_time of J_n^v and the start_time of J_{n+1}^v
- The co-ordinates of the end location of J_n^v were retrieved, i.e., end_lat_v and end_lng_v
- Vehicle availability, a_v, for period p was calculated using Equation (2)
- Each half-hour period, hh_i^d, for which all 30 min fell within p was added to set hh_p^v

- Where $a_v = 1$, the vehicle was deemed to be available for each period within hh_p^v

The resulting dataset contained 677,280 rows, 57% of which represented half-hour periods in which a vehicle was available, i.e., $a_v = 1$. In addition to vehicle availability, several other features were added to the data that had the potential to impact vehicle usage and hence availability:

- The day number (d); from 0 to 6, i.e., Sunday to Saturday
- Half-hour (hh); the index of the half-hour period from 1 to 48
- Public holidays (ph); i.e., national holidays
- University holidays (uh); other days—when the University was closed—that were typically contiguous to public holidays
- Holidays (hol); days that were either a public holiday or a University holiday
- Term days (term), i.e., whether the day fell within a University term period

Example entries in the dataset are shown in Table 2.

Table 2. Sample data from the processed dataset.

Vehicle (v)	Department	d	hh	ph	uh	hol	term	a_v
1	A	1	1	0	1	1	0	0
2	B	3	35	0	0	0	1	1
3	C	5	26	1	0	1	0	0
3	C	5	27	1	0	1	0	1
4	D	4	20	0	0	0	1	1

Where d = day; hh = half-hour period; ph=public holiday; uh = university holiday; hol = holidays; term = university term and a_v = vehicle availability.

The data was split into training and test datasets containing 237 days (81%) and 57 days (19%) of the total dataset, respectively, with the composition shown in Table 3.

Table 3. Composition of the training and test datasets.

Feature	Training	Test
Availability ($a_v = 1$)	57.4%	55.7%
Term days (term = 1)	54.4% (129 days)	47.4% (27 days)
Public Holiday (ph = 1)	3.0% (7 days)	1.8% (1 day)
University Holiday (uh = 1)	1.7% (1 day)	1.8% (1 day)

2.2. Learning Approaches

The learning task was defined as a classification problem. For each half-hour period, the model was tasked with learning and predicting whether the vehicle was available ($a_v = 1$), given the other data as input. The three different approaches used to address this task are described below.

2.2.1. Automated Machine Learning

Successful application of machine learning is critically dependent on the choices made before the learning algorithm is executed. These include the specific algorithm to use for a given problem, how to pre-process the features in the dataset, and how to set the hyperparameters, i.e., the non-optimised configuration of the chosen algorithm. Finding a successful framework is often an iterative and time-consuming process, requiring the training and evaluation of many different algorithms and hyperparameters, which may make the technology inaccessible for non-specialists. These difficulties have led to the development of automated machine learning that typically utilises Bayesian optimisation to search the space of frameworks with the aim of producing an optimised model for the task at hand [19,20]. This simplifies the machine learning workflow and allows the evaluation of a range of proven techniques and implementations for a given problem. This approach has great potential in broadening the use of machine learning and allowing non-specialists in fields, such as energy, to make use of the technology. In this work, two different implementations of this technique were explored:

1. AutoML on Microsoft Azure [21]: At the time of writing, this implementation supported the automated evaluation of up to 16 different algorithms for classification problems, including variations of popular approaches, such as decision trees and gradient boosting. Accuracy was chosen as the primary metric for the optimiser, i.e., the percentage of the training dataset for which availability was correctly predicted, and a typical AutoML run evaluated around 100 different frameworks to produce the final optimised classifier. For the problem explored in this work, the eXtreme Gradient Boosting (XGBoost) classifier was consistently the best performer [22]. This approach is based on gradient boosted decision trees, which is a fast and efficient technique that creates a strong classifier from an ensemble of weak decision tree classifiers.
2. AutoML Tables on the Google Cloud Platform [23]: In addition to considering standard machine learning algorithms, this technique also used neural architecture search (NAS) [24] to assess the efficacy of artificial neural networks. As for other types of machine learning, design of an appropriate neural network for a given problem often requires much trial and error with the number of hidden layers, the number of nodes within each layer, network connectivity and other hyperparameters being key decisions. Best results were achieved by the adaptive structural learning of artificial neural Networks (AdaNet) technique, which progressively builds a network architecture form an ensemble of subnetworks [25].

The results produced by both implementations were not significantly different and, therefore, only one was reported on in this paper. The AdaNet model was chosen as it provided easier access to probabilistic outputs, which were used in the subsequent analysis.

2.2.2. Cumulative Moving Average

Observation of the fleet data suggested a relatively regular pattern of vehicle activity during a typical week. A simple cumulative moving average (CMA) was, therefore, calculated to represent the probability of each vehicle's availability during each half-hour period. Each row of the training dataset was processed to determine the vehicle (v), day (d), half-hour period (hh) and availability (a_v). The corresponding probability (CMA) was then updated using Equation (3). This resulted in 336 probabilities for each vehicle: 48 for each of the 7 days of the week.

$$CMA_n(v,d,hh) = \frac{CMA_{n-1}(v,d,hh) * (n-1) + a_v}{n} \tag{3}$$

A vehicle was predicted to be available, i.e., $a_v = 1$, for a given half-hour period if the associated CMA was greater than 0.5.

2.2.3. Exponential Moving Average

The CMA weights all the data points for each half-hour period equally regardless of how long ago they were received. For static fleet behaviour, this approach may be appropriate; however, in many cases, there are likely to be the changes in how vehicles within the fleet operate over time. The CMA would be slow to adapt to any such changes, which would be of concern for averages constructed over a significant period and thus representing large sets of data points. One method to combat this issue is to use an exponential moving average (EMA) in which the weighting of historical data points decays over time, and more recent data points have a greater influence on the current average, as shown in Equation (4), for a vehicle v in the half-hour period defined by d and hh.

$$EMA_n(v,d,hh) = (a_v - EMA_{n-1}(v,d,hh)) * \left(\frac{2}{N+1}\right) + EMA_{n-1}(v,d,hh) \tag{4}$$

The parameter N determines the weighting given to the most recent data point, a setting of $N = 1$ applies a 100% weighting, whereas larger values of N reduce the weighting. In this work, a value of $N = 20$ was used, thus applying a 9.52% weighting to the most recent data point. It should be noted that this increased weighting in comparison to CMA (for averages of 10 or more data points) also has a potentially negative consequence in emphasising outliers in the data that are not representative of sustained changes in behaviour. As for CMA, a vehicle was predicted to be available for a given half-hour period if the associated average was greater than 0.5.

3. Results and Discussion

In this section, comparative results are presented for the three models during training and on the test dataset. The underlying cause of differences in these results was analysed, and modifications were made to the averaging approaches to account for these differences. The ability of the models to predict availability on a vehicle-by-vehicle basis was then assessed, and a metric was developed to identify vehicles that were good candidates for V2G. Results for cumulative, fleet-level prediction were also presented. The section concludes with a detailed discussion.

3.1. Model Analysis

The training dataset was used to train models using each of the three approaches. Figure 1 shows the confusion matrices and accuracies, following training on the 34-week training dataset. All three models produced similar accuracies with AutoML, achieving a small increase in accuracy over the two averaging approaches. All the models showed a slightly increased propensity for misclassifying periods the vehicle was not available (true label=0) as periods the vehicle was available (predicted label=1), as shown in the upper right quadrants of the confusion matrices.

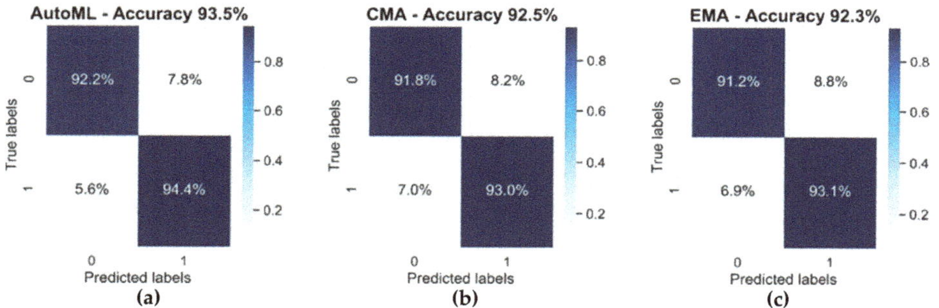

Figure 1. Confusion matrices and accuracy of all 3 models on the training dataset. The labels indicate predicted and actual vehicle availability. AutoML, automated machine learning (**a**); CMA, cumulative moving average (**b**); EMA, exponential moving average (**c**).

To determine if this performance carried over to novel data, the models were tested against the 8-week test dataset. The results in Figure 2 showed that although the accuracy of all 3 models reduced on the test set, the performance remained relatively robust with an accuracy of approximately 90% in all cases. A McNemar test [26] was performed to test the statistical significance between each of the models. The difference between all models was found to be highly significant ($p < 0.001$). This indicated that although the overall accuracy was similar for all models, the set of classification errors made by each approach was significantly different.

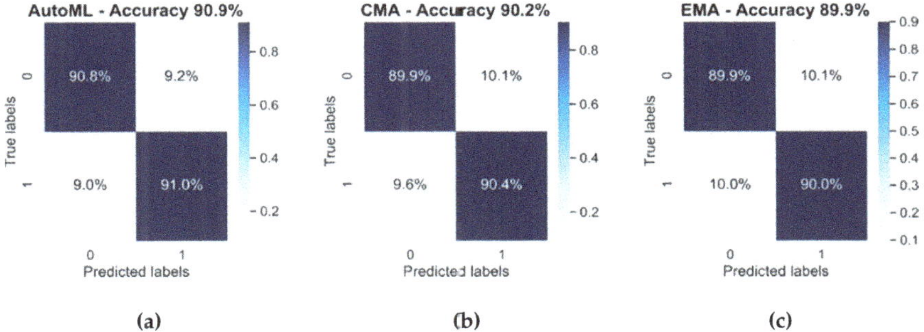

Figure 2. Confusion matrices and accuracies of the 3 models on the test dataset. The labels indicate predicted and actual vehicle availability. AutoML, automated machine learning (**a**); CMA, cumulative moving average (**b**); EMA, exponential moving average (**c**).

To further explore the differences between the models, accuracy was calculated for University term and non-term periods and separately for holidays and non-holidays. Figure 3 shows that performance for term and non-term periods was very similar for all models, including the averaging approaches for which term was not considered during training. This suggested that fleet behaviour was not substantially impacted by this feature. This was not the case, however, for holidays for which the averaging approaches performed poorly and AutoML very well.

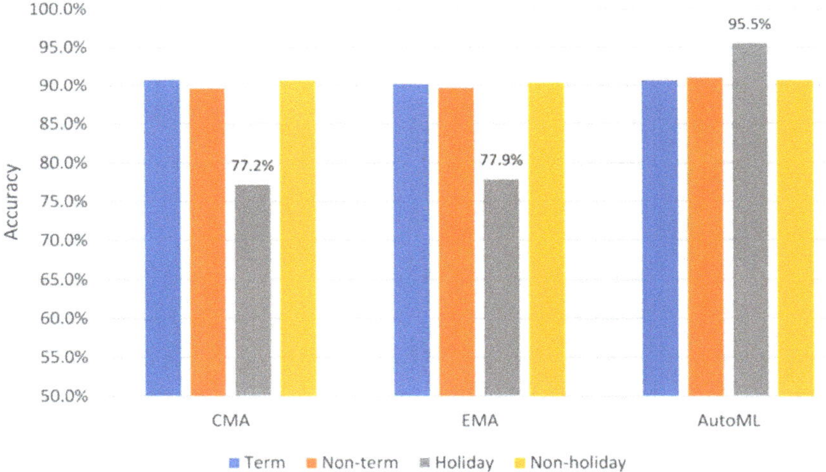

Figure 3. Accuracy of the 3 models for term/non-term periods and holiday/workdays. Data labels show accuracy for holidays.

To demonstrate the reasons for this disparity, the average available vehicles for each half-hour period during the two holidays on Mondays in the test dataset was compared to that predicted by each of the three models. Figure 4 shows that the actual availability was relatively static throughout the holidays, a pattern that was typical for a weekend AutoML correctly identified this pattern; however, the predictions for both CMA and EMA were representative of a typical non-holiday Monday. This was as would be expected, given that the holiday feature was not considered in those approaches.

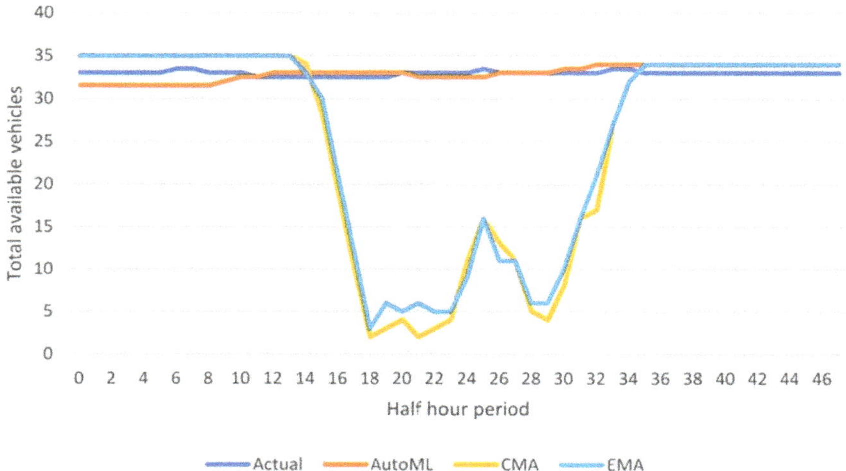

Figure 4. Total available vehicles predicted by the 3 models for the 2 holiday Mondays in the test dataset compared to actual availability.

To accommodate this prediction error, a heuristic was used for the CMA and EMA models that treated any holiday as a Sunday. Therefore, for any rows in the test set with *hol* = 1, the prediction for Sunday was used, i.e., *d* was set to 0 for that row. The revised models, termed CMAh and EMAh, were tested on the same 8 weeks training set using this heuristic and the revised confusion matrices and accuracies, as shown in Figure 5.

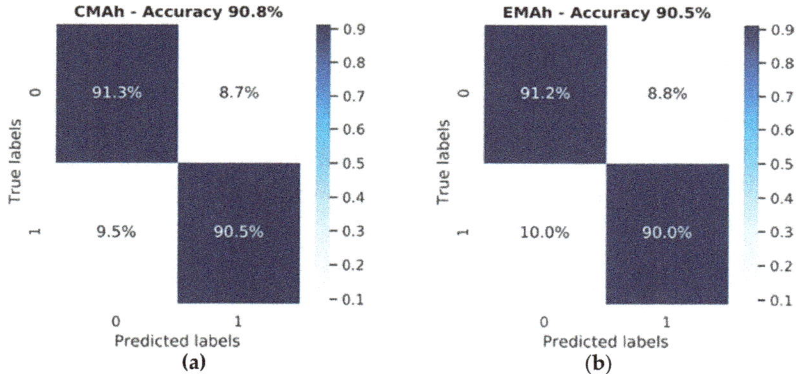

Figure 5. Confusion matrices and accuracies for the averaging approaches using the holiday heuristic. CMAh, cumulative moving average with holiday heuristic (**a**); EMAh, exponential moving average with holiday heuristic (**b**).

The accuracy of both averaging approaches increased through the use of the holiday heuristic and was now comparable to AutoML. A McNemar test was performed and again showed a highly significant difference between the averaging models and the AutoML model ($p < 0.001$). However, no significant difference was now found between the CMAh and EMAh models ($p > 0.05$).

3.2. Vehicle Analysis

To help better understand the performance of the models, the results for each individual vehicle were analysed. Prediction errors for each of the 48 vehicles were determined by calculating the proportion of the test dataset for which the predicted availability was incorrect for that vehicle. Given that the results from the two averaging approaches were not significantly different, the results were only reported for one of the two models (CMAh) for purposes of clarity.

Figure 6 reveals a high degree of correlation between the two models and a clear outlier with a much higher error rate than other vehicles. The analysis of the datasets revealed that was due to a substantially different pattern of behaviour in the 8 test weeks to the 34 training weeks. The vehicle was available for 50.1% of the training period in contrast to only 12.7% of the test period. A similar, but smaller, disparity between training data and test data was also apparent for the two next worse performers. However, this was not always the case for vehicles with a relatively high prediction error. There was a close correlation between the proportion of available periods in the training and test data for the vehicle with the 4th highest error rate despite a prediction error in excess of 20%. In this case, the error was more strongly influenced by the specific times the vehicle was available rather than the total time it was available.

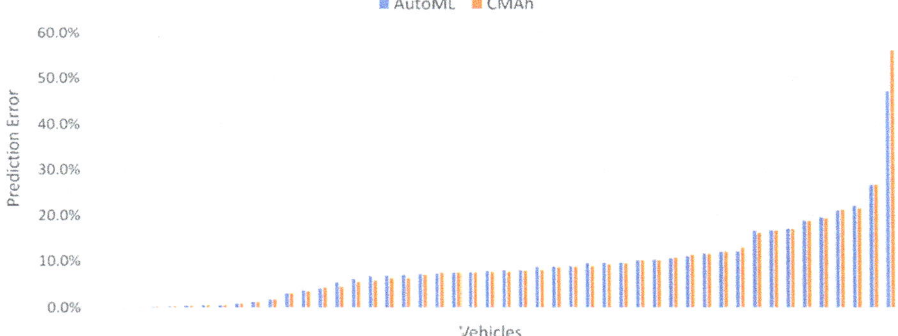

Figure 6. Prediction errors (Perr) for each of the 48 vehicles for the AutoML and CMAh models.

At the other extreme, the figure showed 11 vehicles with error rates of less than 2%. However, the analysis revealed that this excellent performance was enabled by the fact they were almost always unavailable. As a result, both models predicted that these vehicles were never available, and the error rate was due to the small number of periods where this wasn't the case. Such vehicles would not be appropriate for V2G as they must be both relatively predictable and available for substantial amounts of time. A simple metric was thus developed to calculate the viability of a vehicle, given these variables, as shown in Equation (5), where $Perr$ is the prediction error, and Pa_v the percentage of time the vehicle was available, both expressed as a number between 0 and 1.

$$V2Gv = (1 - Perr) * Pa_v \tag{5}$$

Thus, a stationary vehicle that was entirely predictable and always available would score 1. A vehicle that was either entirely unpredictable and/or never available would score 0, and potentially viable vehicles would score somewhere in between. Figure 7 shows this metric calculated for all vehicles using the test dataset and prediction errors from the AutoML model.

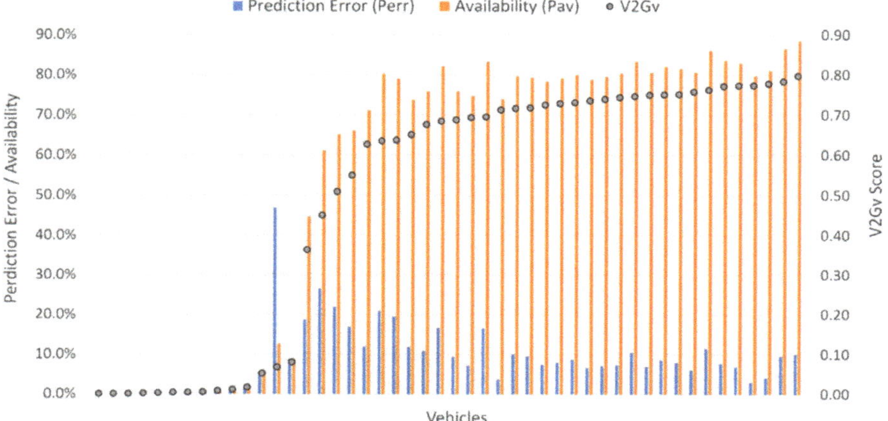

Figure 7. V2G (vehicle-to-grid) viability scores, V2Gv, for the 48 fleet vehicles using the AutoML model on the test dataset.

The figure suggested which vehicles would be candidates for V2G. For example, 30 of the 48 vehicles had a V2Gv score in excess of 0.6 as a result of a combination of relatively low prediction errors and relatively high availability. The same set of 30 vehicles was produced using all three models and consisted of vehicles from every department. Of particular interest for a V2G service, however, is the ability to deliver grid services when most required, i.e., at time of peak demand. To determine whether this was the case, the analysis was repeated, considering only periods within a typical peak demand period of 16:00 to 19:00. Figure 8 shows that 30 vehicles again achieved a V2Gv score over 0.6, with only 1 vehicle differing from the original set. However, 15 vehicles now scored over 0.85, making them excellent candidates for participation in V2G during peak hours.

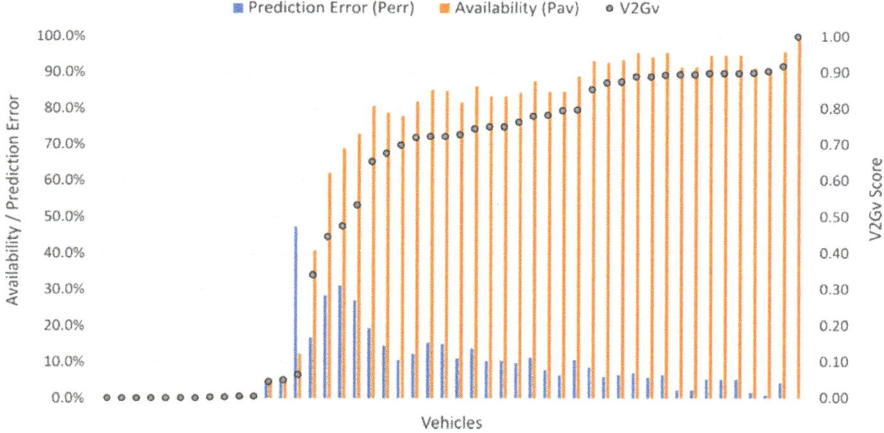

Figure 8. V2Gv scores for the peak 4 pm to 7 pm period using the AutoML model on the test dataset.

Such a score is not in itself sufficient to demonstrate the viability of a vehicle for V2G however. Another key consideration is the ability of the vehicle to deliver the required power or energy when called upon, i.e., it must have sufficient charge to satisfy journey requirements while delivering energy for the V2G service. To assess this requirement, vehicle trip journey over the 34 weeks of training data was analysed to determine the mean daily mileage for each vehicle on a workday and non-workday. This gave an indication of how much battery capacity would be required to satisfy typical journey requirements and hence how much would be available for V2G. The mean workday daily mileage for vehicles with a peak period V2Gv score over 0.6 was found to be only 26 km (s = 21.8 km), and they were rarely used on other days. It would, therefore, be possible to satisfy these journey requirements while enabling V2G with relatively modest battery capacity. In addition, the vehicles were available on average 96.9% (s = 3.9%) of the time, during the hours of 7 pm and 7 am, thus providing the opportunity for them to start the working day fully charged.

3.3. Fleet Analysis

The vehicle analysis presented in the previous section was of value in assessing the viability of individual vehicles for V2G; however, in order to participate in grid services, the pooled available capacity is likely to be of principal concern to an aggregator. One key requirement for predicting this capacity is predicting the total number of vehicles available at a future time, i.e., it may not be necessary to predict the availability of individual vehicles if the total number available can be predicted. Two approaches were used to make this prediction: a sum of individual vehicle's predicted binary availability (SoV) and a sum of individual vehicle's probability of availability (SoP).

To calculate SoV, the binary availability of each vehicle (a_v) predicted by the model (m) for a unique half-hour period (hh_u) in the test dataset was summed, i.e., $total_{a_v}(m, hh_u) = \sum_{v=1}^{n} a_v(m, hh_u)$. The actual number of available vehicles for a period hh_u in the test set was also determined. An error score, $error(m)$, was then calculated for the model m by averaging the percentage error between actual and predicted total availability over all 2736 (57 days * 48) unique half-hour periods in the test dataset. The accuracy of the model was defined as $accuracy(m) = 1 - error(m)$.

The SoP approach was identical to the SoV approach with the exception that the total availability predicted by the model m for half-hour period hh_u was calculated by summing the predicted probability of each vehicle being available, i.e., a threshold was not used to make a binary prediction for each vehicle before summing. For example, given four vehicles, each with a probability of 0.25 that would individually be predicted to be unavailable, this method would predict one vehicle of the group to be available. In this way, vehicles always contributed to the predicted total in correlation with their likelihood of availability.

These calculations were performed for the CMAh and AutoML models. The results, shown in Figure 9, revealed that the accuracy for both models was relatively low using the SoV approach, and no significant difference was found between the two models using a Welch's t-test ($p > 0.05$). However, the use of the SoP approach improved accuracy by 8.2% for CMAh and 9.5% for AutoML, both of which were found to be highly statistically significant improvements ($p < 0.001$). The accuracy of AutoML-SoV was 1.7% higher than that of CMAh-SoV, a result that was also highly statistically significant ($p < 0.001$).

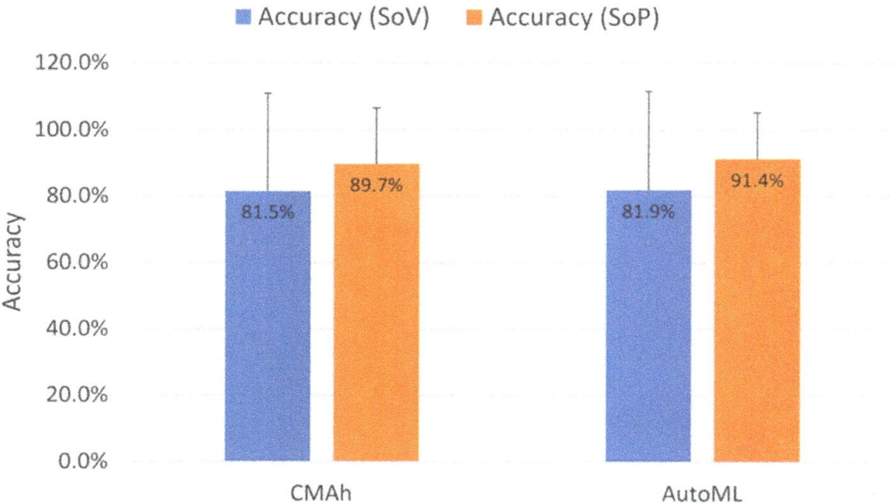

Figure 9. Accuracy of the predicted total number of vehicles over the test period using 2 different approaches (SoV and SoP) for the CMAh and AutoML models (see text for details). Error bars show +1 standard deviation. SoV, the sum of individual vehicle's predicted binary availability; SoP, the sum of individual vehicle's probability of availability.

3.4. Discussion

The learning approaches explored in this work ranged from the simplest averaging techniques to complex machine learning models. However, their performance on the defined task was comparable. This relative equality could be explained by examining the nature of the dataset and the potential patterns of vehicle behaviour that the machine learning approaches had the potential to learn. A University tends to work on annual patterns as it moves through the various terms and holidays. However, as the training set was exclusively drawn from a single year, any annual patterns could not be learned by the machine learning models. This left two other key features that could potentially be utilised, term and holiday. The CMA and EMA averaging techniques did not consider the term, and yet their performance was equivalent during both periods, and thus this feature had little impact on overall vehicle behaviour. In contrast, the holiday feature did impact vehicle behaviour, which was successfully learned by the AutoML model, resulting in improved performance over the averaging techniques. However, the impact was clear and consistent, and, therefore, a simple heuristic was sufficient to compensate for it within the CMA and EMA models.

There was little scope, therefore, for the machine learning approaches to improve over the simple averaging techniques. This, however, would not always be the case. There are many other features that have the potential to impact vehicle behaviour. For the University fleet, these include University open days, special events, weather events and local traffic conditions. Creation of a successful predictive model for vehicle availability is thus not likely to be a one-off event but rather an iterative process where initially available data is used to produce a first model iteration that is retrained and updated as new data becomes available and its performance analysed. For example, observation of periods of significant deviation between actual and predicted availability may allow the identification of events that need to be accommodated within the model. For the examples above, new features may be added to the dataset to identify open days and special events, allowing any associated impact on vehicle behaviour to be learned. Links to live weather and traffic services may also be established so that the impact of various conditions can be accommodated in the data and influence the predictions that are made. As the complexity of the feature set grows and these features interact non-linearly, the impact of individual features will be less easily identifiable, and, therefore, attempting to accommodate them through use of heuristics in the averaging approaches will quickly become untenable. Machine learning approaches can more easily accommodate such complexity and are, therefore, likely to outperform the averaging approaches as a V2G service develops. However, this will require defining the features that have an impact on vehicle predictability and discovering where the relevant data can be found (e.g., labelling of workplace-specific holidays may require parsing events from work calendars, manual input from fleet owners, etc.).

The need to continually iterate and refine the models is also required to enable adaptation to changes in vehicle behaviour. Although the behaviour of the fleet considered in this work was relatively regular, changes would occur over time in response to changes in the way the broader organisation operates, for example. Such changes in schedule would also be apparent for non-fleet users, where they might be more pronounced given that there is likely to be greater flexibility in drivers' schedules. The EMA model used in this work weighted recent data more strongly than historical data to help adapt to changes. However, online or continual learning would also be required for machine learning models to adapt to such concept drift [27].

Analysis of individual vehicles allowed the identification of candidate vehicles for V2G. A "sweet spot" of vehicles was identified that satisfied several enabling requirements: (a) they were available, i.e., parked next to a charge point for a significant amount of time; (b) they were predictable, i.e., errors were low; (c) average daily mileage requirements were relatively low, thus providing spare capacity; (d) they were stationary for at least one extended period, thus allowing the battery to be replenished. Such analysis is of value to a fleet that is considering moving to electric vehicles and the use of V2G services by supporting the prioritisation of vehicles to transition and informing the required capacity of batteries, for example. It is also of importance to assess the number of charge points that are required in each proposed location; even if parked locations can be reliably predicted, this is of little value if all the vehicles cannot find a compatible grid connection. Knowledge of individual vehicles is also important during the operation of a V2G service. It may not be possible to assume the use of a vehicle even if it is plugged in and available as it may be necessary for individuals to receive and accept offers to participate in a given V2G opportunity [28], an issue that may be particularly pertinent for non-fleet users. For non-homogeneous populations of vehicles and batteries, it may also be necessary to target users based on the specific capabilities of their vehicles, such as battery capacity. Such socio-technical considerations have not been widely considered in work to date [29], and more research is required.

In many cases, it will be more beneficial to consider the population of available vehicles rather than individual vehicles. The analysis conducted in this paper showed that considering the cumulative likelihood of vehicle availability was more accurate than making predictions for each vehicle individually, which was especially the case for AutoML. To participate in grid services, the most important thing an aggregator needs to predict is the total capacity available to it at a given time, and the specific vehicles contributing to that capacity may be of lesser concern. However, there are a number of other factors that must be considered when translating vehicle availability to actual available capacity. Chief among these is the battery state of charge, which must be sufficient to enable V2G services while allowing a vehicle to continue operating in its primary role as a form of transport. In this work, the average daily mileage was calculated, which allowed likely available surplus capacity to be assessed for given battery capacity. Such high-level analysis may broadly enable a V2G service; however, more detailed analysis of the historical state of charge and incorporation of such data into the learning algorithm would be of great value to optimise the service. This is particularly true for vehicle populations with larger or less consistent daily mileage, where the explicit state of charge data may be essential to calculating whether a vehicle can participate in a V2G event while retaining enough charge for its next journey. As V2G services develop, such data will be generated as vehicles plug into compatible charge points, which can be used to further refine the models and enable finer-grained capacity predictions.

4. Conclusions

In this work, we compared the use of automated machine learning and moving averages to predict the parked locations of vehicles from a University fleet and their proximity to six proposed sites of V2G charging stations. This allowed the potential availability of vehicles during future half-hour trading periods to be assessed. Prediction errors for individual vehicles were found to be very similar for the simplest averaging techniques and the most complex machine learning techniques. However, this was only enabled using a heuristic for the averaging approaches to adjust for the impact of a key feature in the dataset. This impact was learned without intervention by the AutoML approach, a capability that is of critical importance as the feature set grows and interacts non-linearly making the use of heuristics untenable. Two approaches for using the predictions for individual vehicles to predict the total number of available vehicles were also investigated. It was found that calculating the cumulative probability was more powerful than summing individual vehicle predictions and that AutoML was the most accurate using this approach with an accuracy of 91.4% on the test dataset. While this predictive capability would be of value to a V2G aggregation service, translating available vehicles to available capacity requires the incorporation of other factors, including the state of charge of the battery, which will be a focus of future work.

Author Contributions: Conceptualization, R.S., S.N. and J.P.; data curation, R.S., J.W. and L.R.; formal analysis, R.S.; funding acquisition, R.S. and M.G.; investigation, R.S.; methodology, R.S.; software, R.S.; validation, R.S.; writing—original draft, R.S.; writing—review and editing, J.W., S.N. and J.P. All authors have read and agreed to the published version of the manuscript.

Funding: This research and the APC were funded by the European Space Agency, grant number 4000120818/17/NL/US.

Acknowledgments: This work is part of a collaborative project with our partners Kearney, Brixworth Technologies and Cenex—the centre of excellence for low carbon and fuel cell technologies.

Conflicts of Interest: The authors declare no conflict of interest.

References

1. Villar, J.; Bessa, R.; Matos, M. Flexibility products and markets: Literature review. *Electr. Power Syst. Res.* **2018**, *154*, 329–340. [CrossRef]
2. Waldron, J.; Rodrigues, L.; Gillott, M.; Naylor, S.; Shipman, R. Towards an electric revolution: a review on vehicle- to-grid, smart charging and user behaviour. In Proceedings of the 18th International Conference on Sustainable Energy Technologies (SET 2019), Kuala Lumpur, Malaysia, 20–22 August 2019; Volume 3, pp. 1–9.
3. Kempton, W.; Tomić, J. Vehicle-to-grid power implementation: From stabilizing the grid to supporting large-scale renewable energy. *J. Power Sources* **2005**, *144*, 280–294. [CrossRef]
4. Nottingham City Council. *Nottingham's 2028 Carbon Neutral Charter*; 2019. Available online: http://documents.nottinghamcity.gov.uk/download/7536 (accessed on 14 April 2020).
5. Pudjianto, D.; Ramsay, C.; Strbac, G. Virtual power plant and system integration of distributed energy resources. *IET Renew. Power Gener.* **2007**, *1*, 10. [CrossRef]
6. Nuvve V2G Technology. Available online: https://nuvve.com/technology/ (accessed on 27 February 2020).
7. Payne, G. *Understanding the True Value of V2G*; Cenex: Loughborough, UK, 2019; p. 62. Available online: https://www.cenex.co.uk/app/uploads/2019/10/True-Value-of-V2G-Report.pdf (accessed on 14 April 2020).
8. Powerloop V2G. Available online: https://www.octopusev.com/powerloop (accessed on 27 February 2020).
9. OVO Vehicle-to-Grid Charger. Available online: https://www.ovoenergy.com/electric-cars/vehicle-to-grid-charger (accessed on 27 February 2020).
10. Bates, J.; Leibling, D. *Spaced Out Perspectives on Parking Policy*; RAC Foundation: London, UK, 2012; p. 118. Available online: https://www.racfoundation.org/assets/rac_foundation/content/downloadables/spaced_out-bates_leibling-jul12.pdf (accessed on 14 April 2020).
11. Zhou, Y.; Cao, S.; Hensen, J.L.M.; Lund, P.D. Energy integration and interaction between buildings and vehicles: A state-of-the-art review. *Renew. Sustain. Energy Rev.* **2019**, *114*, 109337. [CrossRef]

12. Iversen, E.B.; Morales, J.M.; Madsen, H. Optimal charging of an electric vehicle using a Markov decision process. *Appl. Energy* **2014**, *123*, 1–12. [CrossRef]

13. Shimizu, O.; Kawashima, A.; Inagaki, S.; Suzuki, T. Vehicle Fleet Prediction for V2G System - Based on Left to Right Markov Model. In Proceedings of the 4th International Conference on Vehicle Technology and Intelligent Transport Systems, Madeira, Portugal, 16–18 March 2018; pp. 417–422.

14. Hou, Y.; Edara, P. Network Scale Travel Time Prediction using Deep Learning. *Transp. Res. Rec.* **2018**, *2672*, 115–123. [CrossRef]

15. Awan, F.M.; Saleem, Y.; Minerva, R.; Crespi, N. A Comparative Analysis of Machine/Deep Learning Models for Parking Space Availability Prediction. *Sensors* **2020**, *20*, 322. [CrossRef] [PubMed]

16. Trakm8 Limited. *Telematic Solutions*. 2019. Available online: https://static.trakm8.com/static/downloads/trakm8-telematics-solutions-brochure.pdf (accessed on 14 April 2020).

17. Hutter, F.; Kotthoff, L.; Vanschoren, J. *Automated Machine Learning*, 1st ed.; Springer International Publishing: New York, NY, USA, 2019; p. 242. ISBN 9783030053178.

18. Waldron, J.; Rodrigues, L.; Gillott, M.; Naylor, S.; Shipman, R. Decarbonising Our Transport System: User Behaviour Analysis to Assess the Transition to Electric Mobility. In Proceedings of the 35th PLEA conference sustainable architecture and urban design (to appear), A Coruña, Spain, 1–3 September 2020.

19. Fusi, N.; Sheth, R.; Elibol, M. Probabilistic matrix factorization for automated machine learning. In Proceedings of the 32nd Conference on Neural Information Processing Systems (NeurIPS 2018), Montréal, Canada, 3–8 December 2018; pp. 3348–3357.

20. Feurer, M.; Klein, A.; Eggensperger, K.; Springenberg, J.; Blum, M.; Hutter, F. Efficient and Robust Automated Machine Learning. In *Advances in Neural Information Processing Systems*, 28th ed.; Cortes, C., Lawrence, N.D., Lee, D.D., Sugiyama, M., Garnett, R., Eds.; Curran Associates, Inc.: New York, NY, USA, 2015; pp. 2962–2970.

21. Microsoft Corporation. What is Automated Machine Learning? Available online: https://docs.microsoft.com/en-us/azure/machine-learning/concept-automated-ml (accessed on 28 February 2020).

22. Chen, T.; Guestrin, C. XGBoost: A scalable tree boosting system. In Proceedings of the 22nd ACM SIGKDD International Conference on Knowledge Discovery and Data Mining, San Francisco, CA, USA, 13–17 August 2016; pp. 785–794.

23. Google AutoML Tables. Available online: https://cloud.google.com/automl-tables (accessed on 6 February 2020).

24. Zoph, B.; Le, Q.V. Neural architecture search with reinforcement learning. In Proceedings of the 5th International Conference on Learning Representations ICLR 2017, Toulon, France, 24–26 April 2017; pp. 1–6.

25. Cortes, C.; Gonzalvo, X.; Kuznetsov, V.; Mohri, M.; Yang, S. AdaNet: Adaptive Structural Learning of Artificial Neural Networks. In Proceedings of the 34th International Conference on Machine Learning, Sydney, Australia, 6–11 August 2017; Volume 70, pp. 874–883.

26. Dietterich, T.G. Approximate Statistical Tests for Comparing Supervised Classification Learning Algorithms. *Neural Comput.* **1998**, *10*, 1895–1923. [CrossRef] [PubMed]

27. Gama, J.; Žliobaitundefined, I.; Bifet, A.; Pechenizkiy, M.; Bouchachia, A. A Survey on Concept Drift Adaptation. *ACM Comput. Surv.* **2014**, *46*, 1–37. [CrossRef]

28. Shipman, R.; Naylor, S.; Pinchin, J.; Gough, R.; Gillott, M. Learning capacity: predicting user decisions for vehicle-to-grid services. *Energy Informatics* **2019**, *2*, 1–22. [CrossRef]

29. Sovacool, B.K.; Noel, L.; Axsen, J.; Kempton, W. The neglected social dimensions to a vehicle-to-grid (V2G) transition: a critical and systematic review. *Environ. Res. Lett.* **2018**, *13*, 13001. [CrossRef]

Article

An MPC Scheme with Enhanced Active Voltage Vector Region for V2G Inverter

Lie Xia [1,2], Lianghui Xu [1,2], Qingbin Yang [1,2], Feng Yu [3,*] and Shuangqing Zhang [1,2]

[1] State Key Laboratory of Operation and Control of Renewable Energy & Storage Systems, Beijing 100192, China; xialie@epri.sgcc.com.cn (L.X.); xulianghui@epri.sgcc.com.cn (L.X.); yangqingbin@epri.sgcc.com.cn (Q.Y.); zhangshuangqing@epri.sgcc.com.cn (S.Z.)

[2] China Electric Power Research Institute, Beijing 100192, China

[3] School of Electrical Engineering, Nantong University, Nantong 226019, China

[*] Correspondence: yufeng628@ntu.edu.cn; Tel.: +86-13776929955

Received: 2 February 2020; Accepted: 10 March 2020; Published: 12 March 2020

Abstract: In this paper, a model predictive control (MPC) scheme with an enhanced active voltage vector region (AV^{2}R) was developed and implemented to achieve better steady-state performance and lower total harmonic distortion (THD) of the output current for a vehicle-to-grid (V2G) inverter. Firstly, the existing MPC methods conducted with single vector and two vectors during one sampling period were analyzed and the corresponding AV^{2}Rs were elaborated. Secondly, the proposed strategy was investigated, aiming at expanding the AV^{2}R and improving the steady-state performance accordingly. A formal mathematical methodology was studied in terms of duty ratio calculation. Lastly, the proposed method was carried out through experimentation. For comparison, the experimental results of the three mentioned methods were provided as well, proving the effectiveness of the proposed algorithm.

Keywords: MPC; AV^{2}R; V2G; inverter

---

1. Introduction

For the sake of fast development of electric vehicle (EV), pertinent research and studies, such as on motor drive [1], battery charger [2] and vehicle to grid (V2G), have received ever-increasing concern. In particular, the V2G system is increasingly emphasized due to its distinct advantages in terms of both EVs and the grid. With respect to EVs, the function can be expanded through the V2G system, and hence, the cost-effectiveness is increased [3]. Meanwhile, for the power grid, V2G creates some interesting features, e.g., the active and reactive power adjustment, the load balance and the frequency regulation, as well as the improvement of the efficiency, stability and reliability [4–6].

As a mobile energy storage unit, the capacity of a single EV is inevitably limited, such as the 4 kWh of the Toyota Prius plug-in hybrid EV and the 85 kWh of the Tesla Model S [3]. Therefore, with respect to the power grid, an individual V2G operation is insufficient and insufficient. However, research on V2G operation for the single EV is still meaningful to alleviate the undesired influence on the grid, especially the current harmonics. This paper concerns the V2G operation of single EV, as shown in Figure 1a. In essence, a V2G structure of the single EV is an inverter connected with the grid and powered by the EV battery. Generally, the mainstream control strategies of the inverter consist of hysteresis current control (HCC) method and proportional-integral (PI) or proportional resonance (PR) controller based on the pulse width modulation (PWM). Particularly, the classical HCC method introduces non-uniform switching frequency, thereby leading to its limited application [7]. To counter this issue, the hysteresis band was modified to operate a uniform switching frequency, while the updated process will inevitably complicate the calculation [8]. With respect to modulation-based control schemes, PI-PWM schemes are utilized to regulate the output current, voltage or power [9].

To obtain better power factor correction performance under the full range of modulation index, a hybrid modulation scheme was provided in [10], where the conventional space vector PWM (SVPWM) was combined with the virtual-vector-based SVPWM. In addition, it should be noticed that the PR-PWM regulator is an alternative solution in terms of the elimination of the steady-state error [11]. Normally, modulation-based PI or PR control strategies feature a good steady-state performance at expenses of reduced dynamic performance, while the parameters tuning in the PI or PR controller is a time-consuming and complex task as well. Another modulation-related method is the combination of model-based dead-beat and the SVM (DB-SVM) strategy [12]. Unlike PI (or PR)-PWM methods, the reference voltage vector in the DB-SVM scheme is directly determined according to the discretized mathematical model of inverter, leading to a faster dynamic response.

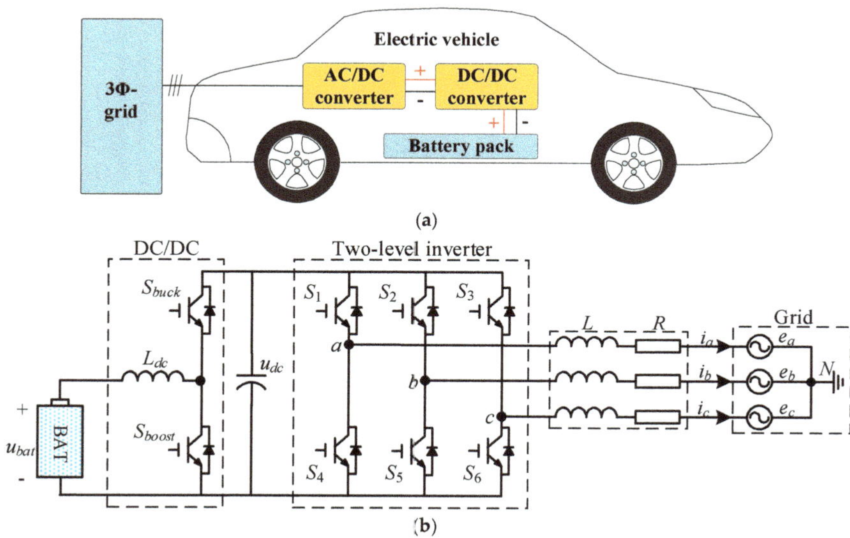

Figure 1. Vehicle to grid (V2G) system for the single vehicle: (**a**) Representational figure; (**b**) The circuit topology.

Recently, the model predictive control (MPC) algorithm has come into the focus of attention since it is characterized by a fast dynamic response, feasible implementation of multi-objective control and strong dynamic decoupling performance [13–21]. It has to be mentioned that the MPC method has some drawbacks, including unfixed switching frequency, lower steady-state performance and undesired computation burden [22], and hence, this limits its implementation in industrial application. To perform the constant switching frequency as well as better steady-state performance, numerous improved MPC methods have been presented. In [23], an MPC using discrete space vector modulation (DSVM) is provided while it yields a higher computation burden [24,25], since 12 vectors must be evaluated during one sampling period. Moreover, a novel MPC scheme based on the space vector modulation (SVM) concept is studied and implemented for a three-level inverter in [26], where the optimum voltage vector is generated through nullifying the derivative of the cost function. In addition, an improved MPC with second modification has been successfully developed for a three-level converter in [27], where the duty-ratios of three vectors are reasonably regulated to achieve a better steady-state performance. In particular, when compared with the DB-SVM method, this scheme can readily realize the multi-objective control by the employment of cost function. Unfortunately, the computation burden is ultra high due to the second optimization for mickle combinations of cost function values [28]. On the whole, these improved arts are to extend the active voltage vector region (AV²R), which will be discussed in Section 2.2, while the computation burden is highly increased, thus complicating the optimization process.

To achieve enhanced AV^{2}R and improved steady-state performance, this paper studies an MPC method in conjunction with SVM concept. Meanwhile, the good dynamic performance of traditional MPC is retained and fixed switching frequency can be acquired in a manner two adjacent active vectors and two null vectors are applied during one sampling period. Significantly, a formal mathematical methodology is conducted in terms of duty ratio calculation for null and active vectors. Moreover, an SVM-based PWM generation method is studied, aiming at reducing the on/off losses and unifying the switching frequency.

This paper is organized as follows: in Section 2, the mathematical model of two-stage grid-connected inverter for V2G is given and analyzed, which consists of a three-phase inverter and a DC/DC converter. Meanwhile, three existing MPC methods for inverter are analyzed and then the AV^{2}Rs in these methods are elaborated. The proposed MPC method that comprises the duty-ratio calculation of the null vector and active vectors determination is conducted in Section 3, where the implementation of AV^{2}R in the proposed method is also discussed. In Section 4, experimental results are examined to verify the effectiveness of the proposed method. Finally, conclusions are presented in Section 5.

2. Modeling and Control of Grid-Connected Inverter for V2G

2.1. Modeling of Two-Stage Grid-Connected Inverter

Referring to Figure 1b, a two-stage grid-connected inverter for V2G is studied in this paper. A DC/DC converter is employed to match the battery voltage and the dc-link voltage, while the two-level inverter converts DC power to AC power absorbed by the grid.

In general, the battery voltage of EV is rated from 300 V to 400 V. However, with respect to the grid-connected inverter, the restriction between dc-link voltage and grid-side voltage must be satisfied as:

$$u_{dc} \geq \sqrt{2}e_{line} \tag{1}$$

where u_{dc} is the dc-link voltage of the inverter; e_{line} is the line-voltage of the grid rated at 380 V as in China.

Due to the bidirectional circuit in the cascaded structure as shown in Figure 1, two operating modes can be initiated, i.e., charging mode and V2G mode. For charging mode, it has to be noticed that the DC/DC part is operated as a buck converter, while the two-level inverter is carried out as a rectifier. In turn, the system is powered by the battery if V2G mode is initiated. In this case, the DC/DC converter operated as a boost circuit upgrades the battery voltage to fulfill the requirement of u_{dc}. Out of the two modes, the V2G operation is targeted for this paper. Concerning the required DC/DC converter, the mathematical model can be derived as:

$$D_{boost} = \frac{u_{dc} - u_{bat}}{u_{dc}} \tag{2}$$

where D_{boost} is the duty cycle of S_{boost}; u_{bat} means the battery voltage.

Next, according to the Kirchhoff voltage law, the mathematical model of the grid-connected inverter can be deduced as:

$$L\frac{\mathrm{d}}{\mathrm{d}t}\begin{bmatrix} i_a \\ i_b \\ i_c \end{bmatrix} = \begin{bmatrix} u_{aN} \\ u_{bN} \\ u_{bN} \end{bmatrix} - R\begin{bmatrix} i_a \\ i_b \\ i_c \end{bmatrix} - \begin{bmatrix} e_b \\ e_b \\ e_c \end{bmatrix} \tag{3}$$

where L is the output filter inductance; R is the line resistance; e_i and i_i ($i = a, b, c$) are the grid voltages and output currents, respectively; u_{iN} is the converter-side voltage, determined by the switching states

S_i. For the sake of conciseness, $S_i = 1$ means the upper switch of phase i is ON, while $S_i = 0$ means the down switch is ON.

Note that (3) is established at each instant, thus, regulation of the output currents can be realized by varying the states of S_i. However, due to the inter-coupling relationship among three phases, the design of the control strategy based on (3) is very demanding. Generally, it is widely acknowledged that the decoupling methods can be implemented at two different coordinate planes, i.e., $\alpha\beta$ and dq frames. In this study, a Park's transformation with invariant amplitude criterion is applied as:

$$T_{abc/dq} = \frac{2}{3}\begin{bmatrix} \cos\theta & \cos(\theta - 2\pi/3) & \cos(\theta + 2\pi/3) \\ -\sin\theta & -\sin(\theta - 2\pi/3) & -\sin(\theta + 2\pi/3) \end{bmatrix} \tag{4}$$

where θ is the angle of the d-axis relative to the a-axis.

Thereafter, the mathematical model of grid-connected inverter in dq frame can be deduced as:

$$\begin{bmatrix} L\frac{di_d}{dt} \\ L\frac{di_q}{dt} \end{bmatrix} = \begin{bmatrix} u_d \\ u_q \end{bmatrix} - \begin{bmatrix} R & \omega L \\ -\omega L & R \end{bmatrix}\begin{bmatrix} i_d \\ i_q \end{bmatrix} - \begin{bmatrix} e_d \\ e_q \end{bmatrix} \tag{5}$$

where the subscripts d and q imply the d-q axis components; ω is the angle frequency of grid voltage.

In essence, the MPC method is a discrete solution to optimize the future behavior of the system, which is applicable in the digital controller. Thus, the continuous model shown in (5) should be transformed to the discrete model. For the sake of simplicity, the future behavior of the control variables is deduced by the forward Euler method as:

$$\begin{bmatrix} L\frac{i_d^{k+1}-i_d^k}{T_s} \\ L\frac{i_q^{k+1}-i_q^k}{T_s} \end{bmatrix} = \begin{bmatrix} u_d^k \\ u_q^k \end{bmatrix} - \begin{bmatrix} R & \omega L \\ -\omega L & R \end{bmatrix}\begin{bmatrix} i_d^k \\ i_q^k \end{bmatrix} - \begin{bmatrix} e_d^k \\ e_q^k \end{bmatrix} \tag{6}$$

where the superscript k represents the kth sampling period; T_s is the sampling period.

2.2. Existing MPC Methods and Corresponding AV²Rs

For a two-level inverter, eight feasible voltage vectors (six active and two null voltage vectors) can be obtained according to the switching states, as presented in Figure 2a. In classical MPC method (termed as MPC1), the system can be simultaneously controlled by evaluating a cost function. With respect to the mentioned V2G inverter, the objective is to regulate the output current $i = [i_d, i_q]^T$. Hence, the cost function can be constructed as:

$$g = \|i^{ref} - i^{k+1}\|^2 \tag{7}$$

Thereafter, eight feasible vectors are utilized to evaluate the cost function and the optimal vector is selected by minimizing the cost function. Nevertheless, since the switching state alternates between 0 and 1, the AV²R (highlighted with red) is insufficient and seven points are merely covered from the graphical subject of view, as observed in Figure 2b. As such, the applied vector is limited to these points. In this manner, the locus of applied vectors is necessarily discontinuous, and therefore, the control effect is also relatively limited.

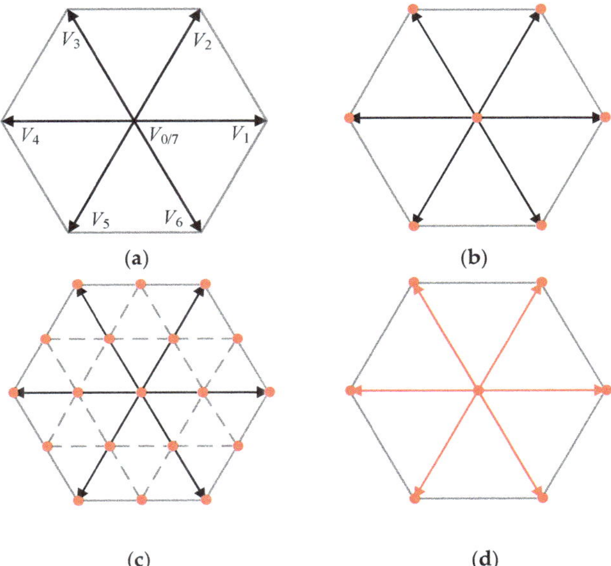

Figure 2. Feasible voltage vectors and active voltage vector region (AV^2R) of existing model predictive control (MPC) methods: (**a**) Feasible voltage vectors; (**b**) AV^2R of MPC1; (**c**) AV^2R of MPC2; (**d**) AV^2R of MPC3.

To this end, an improved strategy using the DSVM method is studied in [23], termed as MPC2. Twelve virtual vectors in conjunction with eight actual vectors are evaluated during each sampling period. Hence, as illustrated in Figure 2c, the AV^2R is extended to 19 points from seven points, which results in that the computation burden is increased extremely. Furthermore, a duty-ratio-based MPC is developed for a three-level inverter in [26], termed MPC3. Carrying out this algorithm in the two-level inverter, AV^2R of six lines can be covered due to that the null vector is inserted as shown in Figure 2d. However, both of these two arts still suffer from the problem of an inadequate AV^2R, and hence, the discontinuous locus of applied vectors.

Comparatively, the art proposed in [27] for a three-level converter is outperforming in terms of the AV^2R. When this method is extended to the two-level inverter, the AV2R will reach the whole hexagon region. Nevertheless, all the actual vectors as well as the vector combinations of the six sectors should be assessed in the two-level inverter, which will introduce an exceptionally high computation burden.

3. Proposed MPC Method

From the aforementioned analysis, although MPC2 and MPC3 schemes can provide a better performance, both of them are weak if they stack up against the conventional PI-SVPWM method, which can yield an AV^2R of the whole hexagon area. In order to address this issue, a four-vector-based MPC is proposed, in which two adjacent active vectors along with two null vectors are applied during each sampling period. In this way, a modulation-like behavior can be realized, yielding the desired steady-state performance and unifying the switching frequency. Meanwhile, it should be mentioned that the excellent dynamic response inherent to the classical MPC can be retained. The control diagram of the proposed MPC is presented in Figure 3. Each part is elaborated as follows.

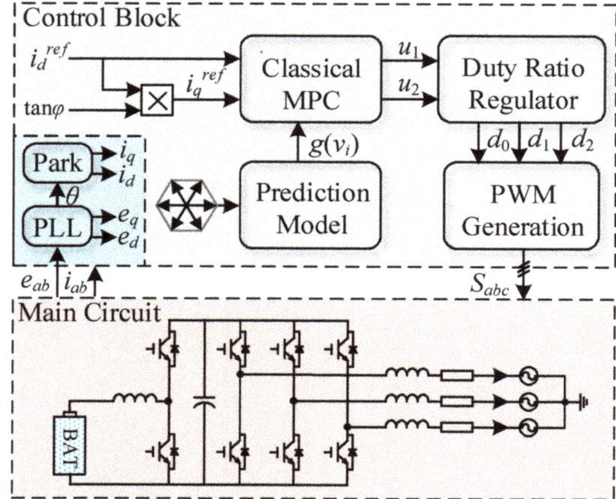

Figure 3. Control block of the proposed MPC method.

3.1. Determination for Two Active Voltage Vectors

In the classical MPC, an optimal vector can be selected through the evaluation of cost function. Based on this concept, two active vectors are selected with a two-stage cascaded algorithm.

The first stage is to determine the optimal vector using conventional MPC. It is noteworthy that the number of voltage vector candidates is six (V_1-V_6) instead of eight. For the sake of convenience, the optimal vector is counted as u_1. Next, the second stage is to decide the suboptimal vector (termed as u_2) from the vectors adjacent to the optimal one. For instance, supposing that the optimal vector is V_1, then V_2 and V_6 are the two adjacent vectors and selected as the candidates associated with the suboptimal vector. Then, if the cost function value generated by V_2 is smaller than that generated by V_6, V_2 will be selected as the suboptimal vector and vice versa. In the same manner, the candidate vectors can be picked out for the other cases (as listed in Table 1), and the suboptimal vector can be determined.

Table 1. Optimal vector and corresponding candidates associated with the suboptimal vector.

Optimal Vector	Candidates Associated with the Suboptimal Vector	
V_1	V_2	V_6
V_2	V_1	V_3
V_3	V_2	V_4
V_4	V_3	V_5
V_5	V_4	V_6
V_6	V_5	V_1

3.2. Duty Ratio Regulator

In fact, according to the traditional PI-SVPWM method, the locus of applied vectors is generally round in shape within the hexagon range illustrated in Figure 2a, where the null vectors are commonly adopted for the better steady-state performance. Therefore, apart from two active vectors determined in Section 3.1, two null vectors V_0 and V_7 (counted as u_0) are inserted during each sampling period as well. This subsection introduces a formal mathematical methodology and targets the calculation of the duration for u_0, u_1 and u_2.

Indeed, the value of cost function (7) expresses the square of current error, termed as ε^2. Since the sampling period is extremely short in general, this error could be mentioned as a linear variable relevant to the duration of the corresponding vector [27]. Hence, considering the case where u_0, u_1 and u_2 are applied and their durations are d_0, d_1 and d_2, the value of cost function can be approximated as:

$$g = \varepsilon^2 = \varepsilon(u_0)^2 d_0^2 + \varepsilon(u_1)^2 d_1^2 + \varepsilon(u_2)^2 d_2^2 \tag{8}$$

where $\varepsilon(u_k)^2$ (k=0, 1 or 2) is equal to the corresponding cost function value generated by u_k.

Note that d_0, d_1 and d_2 subject to the following constraints:

$$\begin{cases} d_0 + d_1 + d_2 = 1 \\ 0 \le d_0 \le 1 \\ 0 \le d_1 \le 1 \\ 0 \le d_2 \le 1 \end{cases} \tag{9}$$

In this manner, the objective is to determine an optimal set of duty ratios with respect to a given (u_0, u_1, u_2), thereby minimizing (8) accordingly. This optimization along with constraints is usually called as the conditional extremum problem, which could be effectively solved by employing the well-noted Lagrange multipliers method in conjunction with the Hessian matrix. By these means, the optimal set of duty ratios, termed as (d_0^*, d_1^*, d_2^*), can be calculated by the utilization of (10) and the Hessian matrix G is provided in (11). Particularly, G is a positive definite matrix with respect to the calculated (d_0^*, d_1^*, d_2^*); namely, this set of duty ratios corresponds to the minimum point of (8). In fact, this problem can be solved using the convex optimization theory and the same result can be deduced as well.

$$\begin{bmatrix} d_0^* \\ d_1^* \\ d_2^* \end{bmatrix} = \begin{bmatrix} \dfrac{\varepsilon(u_1)^2\varepsilon(u_2)^2}{\varepsilon(u_0)^2\varepsilon(u_1)^2+\varepsilon(u_1)^2\varepsilon(u_2)^2+\varepsilon(u_2)^2\varepsilon(u_0)^2} \\ \dfrac{\varepsilon(u_0)^2\varepsilon(u_2)^2}{\varepsilon(u_0)^2\varepsilon(u_1)^2+\varepsilon(u_1)^2\varepsilon(u_2)^2+\varepsilon(u_2)^2\varepsilon(u_0)^2} \\ \dfrac{\varepsilon(u_0)^2\varepsilon(u_1)^2}{\varepsilon(u_0)^2\varepsilon(u_1)^2+\varepsilon(u_1)^2\varepsilon(u_2)^2+\varepsilon(u_2)^2\varepsilon(u_0)^2} \end{bmatrix} \tag{10}$$

$$G(d_0^*, d_1^*, d_2^*) = \begin{bmatrix} \dfrac{\partial^2 g}{\partial d_0^2} & \dfrac{\partial^2 g}{\partial d_0 \partial d_1} & \dfrac{\partial^2 g}{\partial d_0 \partial d_2} \\ \dfrac{\partial^2 g}{\partial d_1 \partial d_0} & \dfrac{\partial^2 g}{\partial d_1^2} & \dfrac{\partial^2 g}{\partial d_1 \partial d_2} \\ \dfrac{\partial^2 g}{\partial d_2 \partial d_0} & \dfrac{\partial^2 g}{\partial d_2 \partial d_1} & \dfrac{\partial^2 g}{\partial d_2^2} \end{bmatrix} \Bigg|_{\substack{d_0 = d_0^* \\ d_1 = d_1^* \\ d_2 = d_2^*}} = \begin{bmatrix} 2\varepsilon(u_0)^2 & 0 & 0 \\ 0 & 2\varepsilon(u_1)^2 & 0 \\ 0 & 0 & 2\varepsilon(u_2)^2 \end{bmatrix} \tag{11}$$

3.3. PWM Generation

For the power electronics and inverter-based machine drive, the PWM generation strategy is the final step of the digital implementation. Since both the proposed method and SVM strategy apply two adjacent active vectors and two null vectors during each sampling period, their PWM generations are essentially the same, as illustrated in Figure 4. For example, assuming that the two selected active vectors are V_1 and V_2, the sequence of vector actions is V_0-V_1-V_2-V_7-V_2-V_1-V_0, as shown in Figure 4a. This operation provides two interesting features simultaneously. On the one hand, just one-phase switching state alters at any instant and reduces the switching loss to some extent. On the other hand, similar to the conventional PI-SVPWM method, the switching frequency is unified since the switching state changes twice during each sampling period. This feature creates an advantage worth mentioning in terms of filter design.

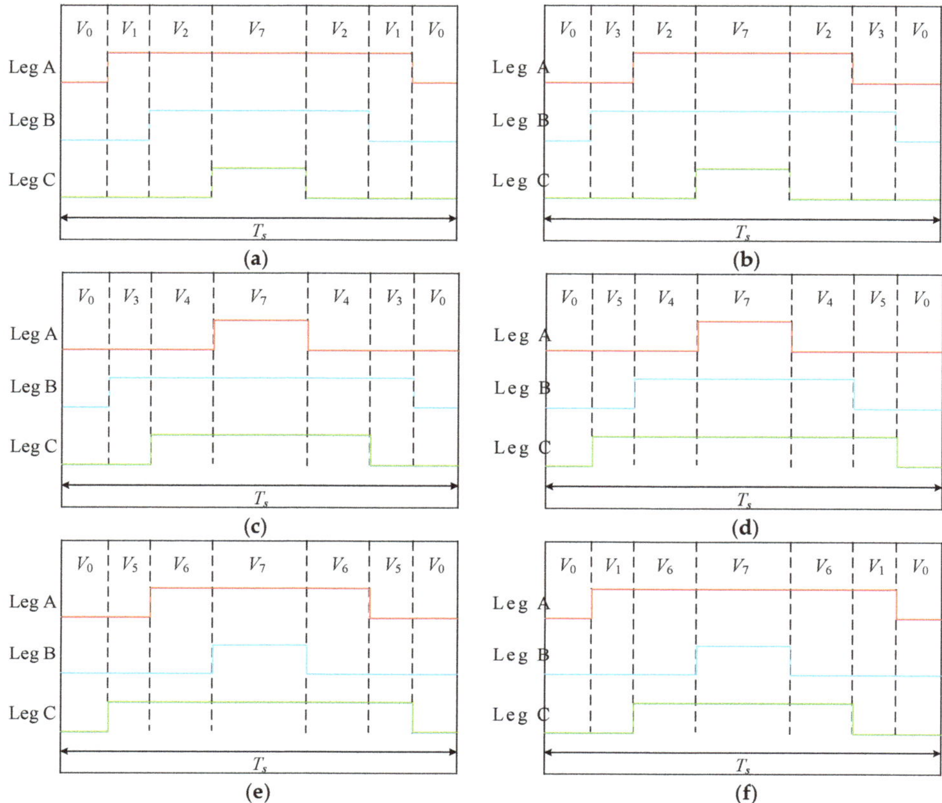

Figure 4. The sequence of vector actions for different vector combinations: (**a**) V_1 and V_2; (**b**) V_2 and V_3; (**c**) V_3 and V_4; (**d**) V_4 and V_5; (**e**) V_5 and V_6; (**f**) V_6 and V_1.

3.4. AV^2R Analysis and Summary of the Proposed MPC

Consequently, the proposed method can enhance the AV^2R and improves the steady-state performance consequently. Note that $(d_0{}^*, d_1{}^*, d_2{}^*)$ is calculated by solving the optimization problem with constraints shown in (10), which ensure $(d_0{}^*, d_1{}^*, d_2{}^*)$ is a feasible solution. Hence, the proposed method can expand the AV^2R to the region of the whole hexagon theoretically, as presented in Figure 5. Compared to three methods of MPC1-3, since all vectors within the hexagon can be generated, the proposed method yields a sufficient AV^2R to create the continuous locus of applied vectors and hence leads to a better steady-state performance.

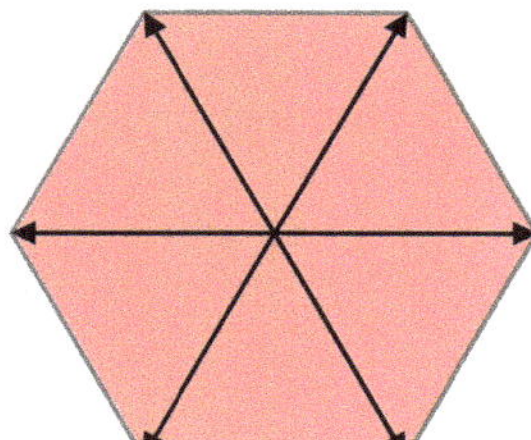

Figure 5. AV^2R of the proposed MPC method.

To sum up, the implementation of the Lagrange multipliers method to determine the duration of the null vector and two active vectors is the core idea of this paper. The process of the proposed algorithm steps can be listed as follows.

Step 1. Measurement of grid-voltage and output currents. Then, to calculate θ, i_d, i_q, e_d and e_q using phase-locked-loop (PLL) and Park's transformation.

Step 2. Determination of two active voltage vectors using MPC1, as discussed in Section 3.1.

Step 3. Duty ratio calculation for null and active vectors, as discussed in Section 3.2.

Step 4. PWM generation based on the SVM concept, as discussed in Section 3.3.

4. Experimental Results

In order to verify the effectiveness of the proposed method, a test prototype of the grid-connected inverter was conducted, as shown in Figure 6. For the AC-side, a 1 kW three-phase transformer was utilized. Both primary and secondary windings of the transformer were Y-connected, while the neutral point was not elicited, and the ratio was 10:1. Meanwhile, three inductors with 5 mH/0.7 Ω were employed as the output filter inductances. For DC-link, an adjustable dc power supply was used to emulate an EV battery and the dc/dc converter was not required consequently. The dc power supply could provide a maximum voltage of 300 V, while the tested voltage was 150 V. Moreover, the three-leg inverter consisted of three FF300R12ME4 (Infineon) modules. Two current sensors WHB25LSP3S1 were utilized to measure the output currents (i_b and i_c). The grid voltages (e_{ab} and e_{bc}) and dc-link voltage were measured using three voltage sensors WHV05AS3S6. Furthermore, a TMS320F28335 digital signal processor was employed to implement the real-time control code, which was developed with C language in Code Composer Studio 6. 0. The sampling frequency was set to 10 kHz. What should be noted is that the grid voltage e_a was exported to the oscilloscope through digital to analog (D/A) channel since the neutral point of the used transformer was not elicited. Here, a D/A chip AD5344BRU was utilized to implement the D/A function.

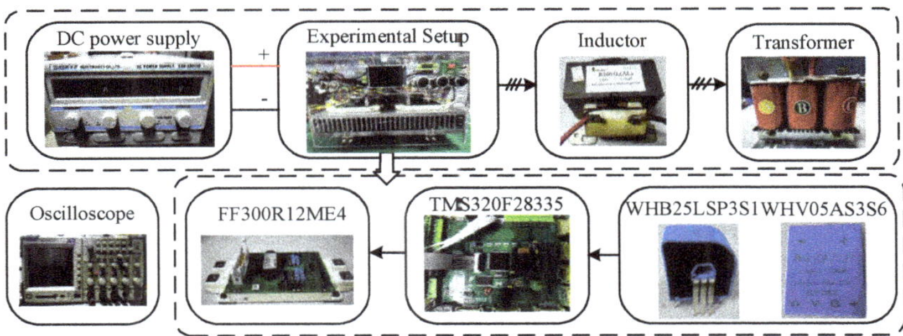

Figure 6. Experimental setup.

4.1. Steady-State Performance Comparison

At first, the steady-state performance comparison was carried out. The d-axis current reference i_d^{ref} was set to 8 A and the power factor angle φ was set to 0. In this case, the inverter was operated under the unity power factor condition. The total harmonic distortion (THD) analysis was conducted using the fast Fourier transform (FFT) tool of the MATLAB/Simulink powergui block.

The results are presented in Figure 7. It can be observed that the four MPC methods yielded sinusoidal output currents and the peak-value of currents was about 8 A. Meanwhile, the unity power factor operation could be realized. However, MPC1, MPC2 and MPC3 provided THDs of 19.73%, 15.68% and 17.28%, respectively. Comparatively, the THD generated with the proposed MPC was within 10%. This extreme improvement was as expected since the proposed MPC can extend the AV^2R to the whole hexagon instead of some points or lines in the existing arts. In other words, the proposed MPC can create a continuous-changed control signal, while the existing methods just created step-changed signals. This comparison proves that the steady-state performance can be greatly enhanced and the tracking accuracy can be realized using the proposed MPC.

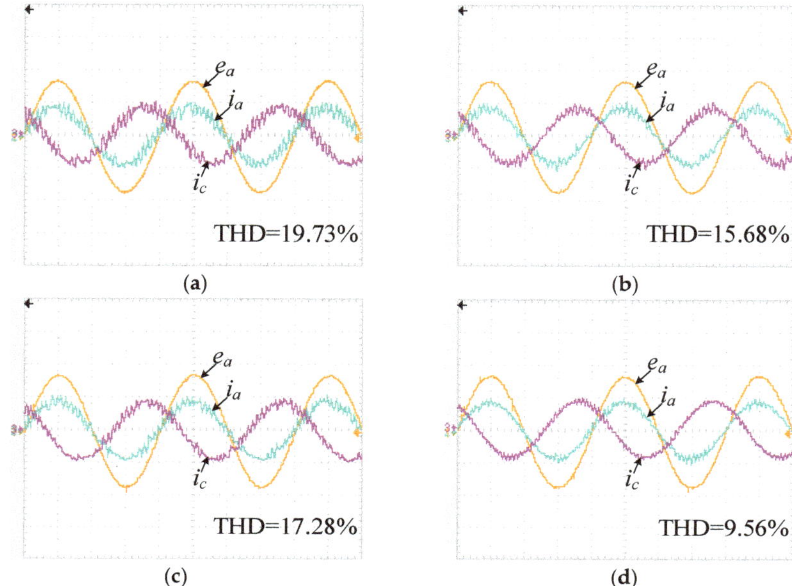

Figure 7. Steady-state performance of four MPC methods (20 V/div, 10 A/div, 5 ms/div): (**a**) MPC1; (**b**) MPC2; (**c**) MPC3; (**d**) Proposed MPC.

4.2. Dynamic Performance Comparison

Next, the *d*-axis current reference step-change test was implemented to investigate the dynamic performance of the proposed MPC method. The *d*-axis current reference i_d^{ref} was suddenly changed from 8 A to 5 A. Meanwhile, it should be noticed that the power factor angle φ was maintained as 0 to ensure the unity power factor operation.

The experimental waveforms are illustrated in Figure 8. Four MPC methods can provide a fast transient of the current response. Meanwhile, the unity power factor operation can also be realized under the post-step-change condition. The peak-value of phase currents was about 8 A and 5 A in the two cases, respectively. This comparison indicates that the proposed art does not sacrifice the dynamic performance to generate a better steady-state performance. Indeed, MPC methods determined an optimal control signal to regulate the control variable and the proposed method provided a novel implementation of optimization process while it retained the excellent dynamic performance of the classic MPC method.

4.3. Performance under Non-Unity Power Factor Conditions

With respect to the V2G system, the adjustment of active and reactive power was an essential task. To evaluate the gird-connection performance under the non-unity power factor conditions, two cases where the power factor angle φ was suddenly changed from 0 to $\pi/6$ and from 0 to $-\pi/6$ were implemented, respectively. The results are presented in Figure 9.

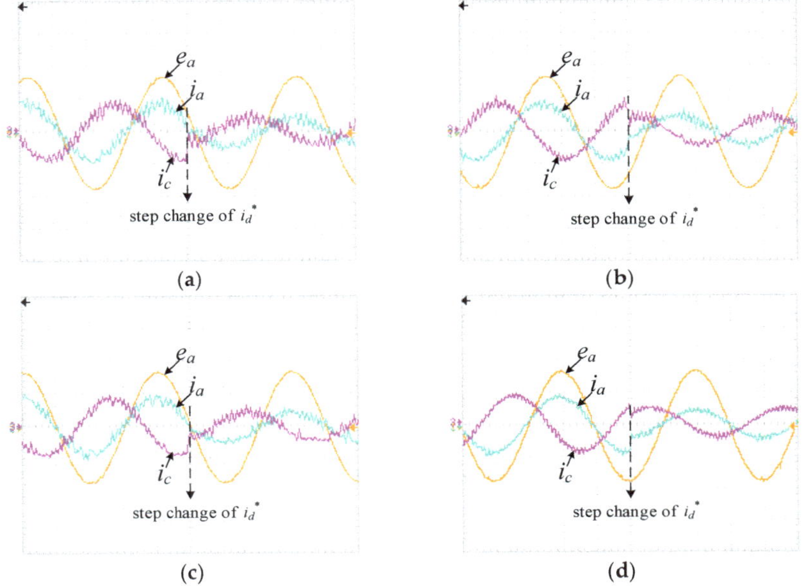

Figure 8. Dynamic performance of four MPC methods (20 V/div, 10 A/div, 5 ms/div): (**a**) MPC1; (**b**) MPC2; (**c**) MPC3; (**d**) Proposed MPC.

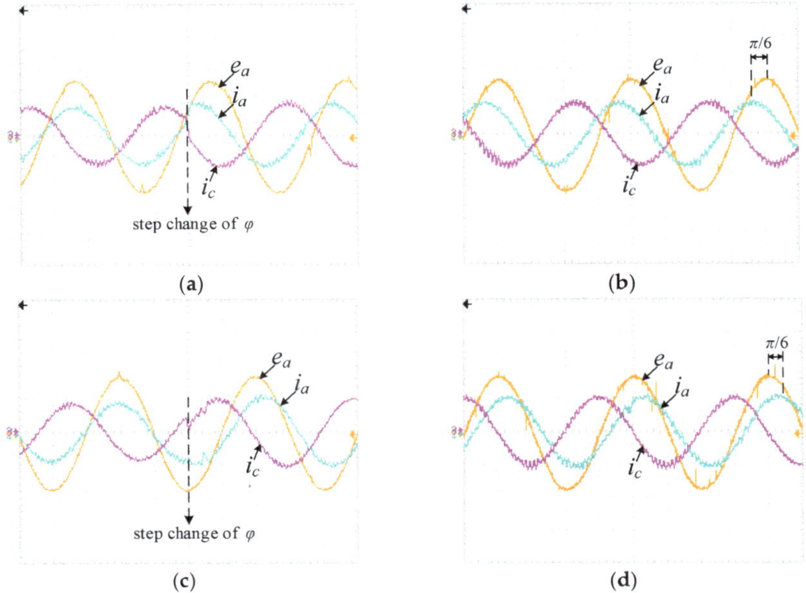

Figure 9. Performance of non-unity power factor conditions (20 V/div, 10 A/div, 5 ms/div): (**a**) Dynamic performance under the condition of φ is suddenly changed from 0 to $\pi/6$; (**b**) Steady-state performance under φ is $\pi/6$; (**c**) Dynamic performance under the condition of φ is suddenly changed from 0 to -$\pi/6$; (**d**) Steady-state performance under φ is -$\pi/6$.

From Figure 9a,c, it can be found that excellent dynamic response can be achieved when φ had a sudden change. In particular, from Figure 9b, the phase current i_a was shifted about $\pi/6$ with respect to e_a and the peak-value of the phase current had a slight increase, which can be explained by the fact that $i_q{}^{ref}$ desires a non-zero value under non-unity power factor condition. A similar result can be seen in Figure 9d, and a shifted angle of $-\pi/6$ of phase current can be generated. These results validated that excellent performance under non-unity power factor conditions can be obtained using the proposed MPC.

4.4. Some Discussions

The obtained results verify that the proposed method can provide better steady-state performance and maintain the inherited dynamic capability. Nevertheless, there are several practice-related issues that should be further discussed.

On the one hand, with respect to the equal voltage value, a DC power supply performs the same electrical characteristic as a healthy battery pack installed onboard the EV when the consideration of battery internal resistance variation is ignored, hence, it is quite feasible to conduct the proof-concept prototype with a DC power supply. It should also be noted that this replacement will be ill-suited when the battery pack suffers from internal resistance variation during a long-term deep discharge process. In practice, the higher internal resistance will lead to a relatively lower output voltage, which may yield an unsatisfactory implementation of the V2G system.

On the other hand, when the grid current has a peak value of 8 A, the voltage loss on the filter inductor will be about 18% (with respect to the voltage of the transformer's secondary side) due to the $0.7\ \Omega$ internal resistance of the filter inductor. To compensate for the voltage loss, the dc-link voltage must be increased, which in turn results in a large current ripple as observed in the experimental results. For the alleviation of the problems of the high dc-link voltage accompanied with the large current ripple, the filter inductor with lower internal resistance can be theoretically adopted. Moreover, the utilization of LCL filter can be regarded as an alternative solution because it holds the definite merit of a better performance in terms of the electromagnetic interference suppression.

5. Conclusions

In this paper, an MPC method with an enhanced AV^2R was proposed for a two-level grid-connected inverter. First of all, three existing MPC methods and the corresponding AV^2Rs were analyzed and discussed. Thereafter, to improve the steady-state performance of the MPC method, a formal mathematical methodology was studied in terms of duty-ratio calculation, in which the well-known Lagrange multipliers method and the Hessian matrix were applied. The AV^2R of the proposed MPC method and an SVM-based PWM generation strategy were elaborated as well. Finally, the experimental results of the proposed MPC were presented. Moreover, to prove the effectiveness of the proposed method, three mentioned MPC methods were conducted. The comparative results verified that the proposed MPC method can improve the steady-state performance and retain the excellent dynamic performance of the classic MPC method simultaneously. Furthermore, the grid-connected operation under non-unity power factor conditions can also be realized using the proposed art, ensuring that the V2G system is characterized with excellent performance under different desired power factor.

Author Contributions: Conceptualization, L.X. (Lie Xia), L.X. (Lianghui Xu), F.Y.; Methodology, L.X. (Lie Xia); Software, L.X. (Lie Xia); Validation, L.X. (Lie Xia), L.X. (Lianghui Xu) and Q.Y.; Formal Analysis, L.X. (Lie Xia) and Q.Y.; Investigation, L.X. (Lie Xia); Resources, L.X. (Lianghui Xu) and Q.Y.; Data Curation, L.X. (Lie Xia) and L.X. (Lianghui Xu); Writing-Original Draft Preparation, L.X. (Lie Xia); Writing-Review & Editing, F.Y. and S.Z.; Visualization, L.X. (Lie Xia), L.X. (Lianghui Xu) and Q.Y.; Supervision, S.Z. Project Administration, S.Z.; Funding Acquisition, Q.Y. All authors have read and agreed to the published version of the manuscript.

Funding: This research was funded by the National Key Research and Development Program of China (2018YFB1500705).

Conflicts of Interest: The authors declare no conflict of interest.

References

1. Cheng, M.; Yu, F.; Chau, K. Dynamic performance evaluation of a nine-phase flux-switching permanent-magnet motor drive with model predictive control. *IEEE Trans. Ind. Electron.* **2016**, *63*, 4539–4549. [CrossRef]
2. Yu, F.; Zhang, W.; Shen, Y. A nine-phase permanent magnet electric-drive-reconstructed onboard charger for electric vehicle. *IEEE Trans. Energy Convers.* **2018**, *33*, 2091–2101. [CrossRef]
3. Chau, K. *Electric Vehicle Machines and Drives: Design, Analysis and Application*; Wiley Press: Hoboken, NJ, USA, 2015.
4. Koyanagi, F.; Uriu, Y. A strategy of load leveling by charging and discharging time control of electric vehicles. *IEEE Trans. Power Syst.* **1998**, *13*, 1179–1184. [CrossRef]
5. Yilmaz, M.; Krein, P.T. Review of the impact of vehicle-to-grid technologies on distribution systems and utility interfaces. *IEEE Trans. Power Electron.* **2013**, *28*, 5673–5689. [CrossRef]
6. Nunna, H.; Battula, S.; Doolla, S. Energy management in smart distribution systems with vehicle-to-grid integrated microgrids. *IEEE Trans. Smart Grid* **2018**, *9*, 4004–4016. [CrossRef]
7. Rahman, K.; Khan, M.; Choudhury, M. Implementation of programmed modulated carrier HCC based on analytical solution for uniform switching of voltage source inverters. *IEEE Trans. Power Electron.* **2003**, *18*, 188–197. [CrossRef]
8. Komurcugil, H.; Bayhan, S.; Abu-Rub, H. Variable- and fixed-switching-frequency-based HCC methods for grid-connected VSI with active damping and zero steady-state error. *IEEE Trans. Ind. Electron.* **2017**, *64*, 7009–7018. [CrossRef]
9. Gonçalves, A.; Aguiar, C.; Bastos, R. Voltage and power control used to stabilise the distributed generation system for stand-alone or grid-connected operation. *IET Power Electron.* **2016**, *9*, 491–501. [CrossRef]
10. Jiang, W.; Du, S.; Chang, L. Hybrid PWM strategy of SVPWM and VSVPWM for NPC three-level voltage-source inverter. *IEEE Trans. Power Electron.* **2010**, *25*, 2607–2619. [CrossRef]
11. Zmood, D.; Holmes, D. Stationary frame current regulation of PWM inverters with zero steady-state error. *IEEE Trans. Power Electron.* **2003**, *18*, 814–822. [CrossRef]
12. Cheng, C.; Nian, H.; Wang, X. Dead-beat predictive direct power control of voltage source inverters with optimised switching patterns. *IET Power Electron.* **2017**, *10*, 1438–1451. [CrossRef]
13. Pedro, G.; Sérgio, C.; André, M. Finite control set model predictive control of six-phase asymmetrical machines—An overview. *Energies* **2019**, *12*, 4693.
14. Wang, W.; Lu, Z.; Hua, W. Simplified model predictive current control of primary permanent-magnet linear motor traction systems for subway applications. *Energies* **2019**, *12*, 4144. [CrossRef]
15. Yu, F.; Liu, X.; Zhang, X. Model predictive virtual-flux control of three-phase Vienna rectifier without voltage sensors. *IEEE Access* **2019**, *7*, 169338–169349. [CrossRef]
16. Formentini, A.; Trentin, A.; Marchesoni, M. Speed finite control set model predictive control of a PMSM fed by matrix converter. *IEEE Trans. Ind. Electron.* **2015**, *62*, 6786–6796. [CrossRef]
17. Siami, M.; Khaburi, D.; Rivera, M. A computationally efficient lookup table based FCS-MPC for PMSM drives fed by matrix converters. *IEEE Trans. Ind. Electron.* **2017**, *64*, 7645–7654. [CrossRef]
18. Daniel, W.; Ryszard, S. Sensorless predictive control of three-phase parallel active filter. In Proceedings of the AFRICON 2007, Windhoek, South Africa, 26–28 September 2007.
19. Krzysztof, K. Model predictive control of five-level ANPC converter with reduced number of calculations. In Proceedings of the 2019 PRECEDE, Quanzhou, China, 31 May–2 June 2019.
20. Abousoufiane, B.; Kamel, K.; Aissa, C. Prediction-based deadbeat control for grid-connected inverter with L-filter and LCL-filter. *Electr. Power Compon. Syst.* **2014**, *42*, 1266–1277.
21. Novak, M.; Dragicevic, T. Supervised imitation learning of finite set model predictive control systems for power electronics. *IEEE Trans. Ind. Electron.* **2020**. [CrossRef]
22. Aguirre, M.; Kouro, S.; Rojas, C. Switching frequency regulation for FCS-MPC based on a period control approach. *IEEE Trans. Ind. Electron.* **2018**, *65*, 5764–5773. [CrossRef]
23. Hassine, I.; Naouar, M.; Mrabet-Bellaaj, N. Model predictive-sliding mode control for three-phase grid-connected converters. *IEEE Trans. Ind. Electron.* **2017**, *64*, 1341–1349. [CrossRef]

24. Vazquez, S.; Leon, J.; Franquelo, L. Model predictive control with constant switching frequency using a discrete space vector modulation with virtual state vectors. In Proceedings of the 2009 IEEE International Conference on Industrial Technology, Gippsland, VIC, Australia, 10–13 February 2009.
25. Alam, K.; Xiao, D.; Akter, M. Modified model predictive control with extended voltage vectors for grid-connected rectifier. *IET Power Electron.* **2018**, *11*, 1926–1936. [CrossRef]
26. Sebaaly, F.; Vahedi, H.; Kanaan, H. Novel current controller based on MPC with fixed switching frequency operation for a grid-tied inverter. *IEEE Trans. Ind. Electron.* **2018**, *65*, 6198–6205. [CrossRef]
27. Donoso, F.; Mora, A.; Cárdenas, R. Finite-Set model-predictive control strategies for a 3L-NPC inverter operating with fixed switching frequency. *IEEE Trans. Ind. Electron.* **2018**, *65*, 3954–3965. [CrossRef]
28. Xiao, D.; Alam, K.; Norambuena, M. Modified modulated model predictive control strategy for a grid-connected converter. *IEEE Trans. Ind. Electron.* **2020**. [CrossRef]

Article

Electric Vehicles Energy Management with V2G/G2V Multifactor Optimization of Smart Grids

Gabriel Antonio Salvatti, Emerson Giovani Carati *, Rafael Cardoso and Jean Patric da Costa and Carlos Marcelo de Oliveira Stein

Universidade Tecnológica Federal do Paraná-UTFPR, Pato Branco-PR 85503-390, Brazil; gsalvatti94@gmail.com.br (G.A.S.); rcardoso@utfpr.edu.br (R.C.); jpcosta@utfpr.edu.br (J.P.d.C.); cmstein@utfpr.edu.br (C.M.d.O.S.)
* Correspondence: emerson@utfpr.edu.br

Received: 5 December 2019; Accepted: 28 January 2020; Published: 5 March 2020

Abstract: Energy Storage Systems (ESS) and Distributed Generation (DG) are topics in a large number of recent research works. Moreover, given the increasing adoption of EVs, high capacity EV batteries can be used as ESS, as most vehicles remain idle for long periods during work or home parking. However, the high EV penetration introduces some issues related to the charging power requirements, thereby increasing the peak demand for microgrids where EV chargers are installed. In addition, photovoltaic distributed generation is becoming another issue to deal with in EV charging microgrids. Therefore, this new scenario requires an Energy Management System (EMS) able to deal with charging demand, as well as with generation intermittency. This paper presents an EMS strategy for Microgrids that contain an EV parking lot (EVM), Photovoltaic (PV) arrays, and dynamic loads connected to the grid considering a Point of Common Coupling (PCC). The EVM-EMS utilizes the projections of future PV generation and future demand to accomplish a dynamic programming technique that optimizes the EVs' charging (G2V) or discharging (V2G) profiles. This algorithm attends to user preferences while reducing the demand grid dependences and improves the microgrid efficiency.

Keywords: smart grid; optimization; energy management; electric vehicles; distributed generation

---

1. Introduction

A microgrid can be seen as a small energy network composed of loads, Distributed Generators (DG), and in some cases, Energy Storage Systems (ESS) to support supply demands [1]. Currently, conventional generation systems are shifting to lower capacity DG. In addition, the application of ESS is increasing and becoming very common. These technologies are significantly changing the energy landscape, spreading the concept of microgrids [2].

The increasing adoption of lithium-ion batteries for use in stationary banks or other applications such as electric vehicles has led to a reduction in the production costs of this technology [2]. Some studies showed that a price reduction of around 50% in lithium-ion batteries used in electric vehicles could occur over the next decade, keeping them at a lower cost than the price paid for stationary batteries' kWh [3]. Therefore, market interest in applications that benefit both the power grid and the environment is growing, which limits energy expenditure and reduces greenhouse gas emissions [4].

The automotive EV market has been increasing over the years, due to the decrease in the cost of its components and the concerns to reduce greenhouse gas emissions [4–9]. These concerns have forced industries and countries to implement plans to reduce the consumption of fossil fuels, such as the European target to have an EV share of 80% by 2050 [10]. However, this goal will add an extra demand of 150 GW in the European power system, resulting in demand peaks arising from EV charging stations [8,10,11]. Hence, although EVs with Photovoltaic renewable sources (PV) can be a suitable environmental and economic option assisting the reduction of fossil fuel dependence, when a large number of units is considered, higher demand peaks may erupt, causing stresses to the power system [4,5]. A typical issue of microgrids with EV charging and PV generation can be seen in Figure 1. The PV power depends on the solar irradiance while load and EV power are user dependent. The power grid results from the balance of EV, load, and PV generation.

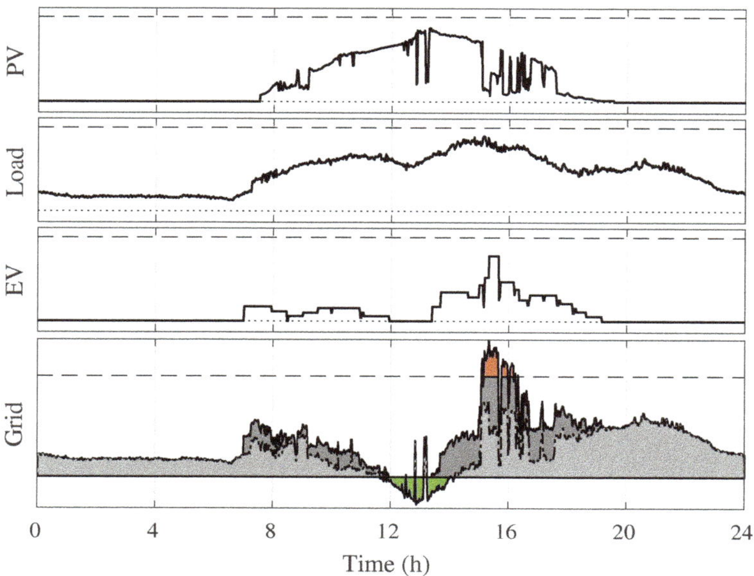

Figure 1. Impact of electric vehicles' charging on the demand of a microgrid.

It is noted that there are periods in which the microgrid demand exceeds the contracted power due to EV charging. In another case, the PV surplus is not efficiently utilized. The results of this analysis point to the need for an Energy Management System (EMS) that properly coordinates EV units in the microgrid and takes into account the occurrence of peak demand. In addition, the EVM-EMS is expected to be able to store surplus energy in EV batteries in order to supply the system when needed (Vehicle to Grid (V2G)). Electric vehicles can act both as an alternative energy storage and as a generator to support the grid, improving efficiency and reliability [1,5,8,9]. Even more, other functions can be added to the EMS to allow ancillary services of frequency and voltage regulation in distribution grids [8]. Since for most of the parking time, the vehicles remain idle, the EVM-EMS enables more flexibility to the demand control [5,11].

If electric vehicles are properly considered in the management system, it may be possible to reduce demand peaks and integrate the elements of micro-grids. This will contribute to economic benefits, while using energy from renewable sources more efficiently and reliably. However, the coordination of intermittent power sources, dynamic loads, and unpredictability of EV, while maximizing efficiency and minimizing costs, implies a large number of obstacles that the EMS must overcome [1,2,9,11].

A significant number of techniques have been proposed for implementing microgrid managers in order to achieve multiple benefits, e.g., maximizing profit, minimizing operation costs, and reducing emissions. In [12], a multivariable strategy was proposed that intended to minimize operation costs while decreasing voltage deviation in the IEEE 33 bus system.

Most of the management systems apply optimization methods and generation and demand forecasting to improve the results of microgrid operation. In [13], irradiance and wind speed predictions were used for PV and wind renewable power sources and a heuristic based algorithm was built in the Python language. This strategy considers energy balance, charging limits, State Of Charge

(SOC), and power generation. Another approach presented by [14] investigated microgrid management considering the energy price, taking into account the energy exchange with external agents. Model Predictive Control (MPC) was employed in the optimization algorithm, which considered the operating cost and the battery degradation in the cost function specification. In [6], the management established by the authors consisted of two parts: a modular topology that used an Autoregressive Integrated Moving Average (ARIMA) model for forecasting and Mixed-Integer Linear Programming (MILP) for optimization.

A two-step management strategy was presented in [7], which separated the load profile calculation from the PV forecast estimate. Moreover, a few works described heuristics optimization techniques. The authors in [15] formulated a game theory to minimize microgrid operation costs or maximize profit from generation units. Additional management methods handled hierarchical control to coordinate the active and reactive power dispatch [16]. The proposed microgrid manager was verified for IEEE 33 buses. Furthermore, stochastic optimization techniques were used for frequency regulation and power management [17], in which the authors followed a mixed strategy of a Bayesian estimator with a Kalman filter to achieve the energy price. Several other papers focused on real-time strategies for optimal power distribution. In [18], the authors applied the MPC method to dispatch power among charging stations. In comparison with the proposed article, a similar EMS strategy was given by [19], who proposed a new EMS approach that took the user preferences and prediction data into account, and the MILP method was chosen for the optimization using software MATLAB/GAMS to solve the computation.

This paper proposes a real-time optimized EMS with a multiple rule decision strategy for microgrids with electric vehicle charging stations. The microgrid is assumed to be comprised of dynamic user loads, PV units operating on the Maximum Power Point (MPP), and an EV parking lot with charging stations wherein all elements are connected to the external grid. The proposed scheme takes the PV generation forecast and the modes priority into account and applies dynamic programming for the optimization process, considering multiple factors in the cost functions. The proposed management system provides four EV charging modes to provide the user several options, as ultra/fast charging or energy/cost efficiency. The ECO and V2G modes apply dynamic programming to optimize the EV battery operation. The proposed approach takes into account the user preferences while aiming to alleviate the microgrid demand and make a profit for the facility owner. Moreover, the EMS system is implemented in a feasible architecture to verify the dynamic programming operation in real-time conditions.

The work is organized as follows: Section 2 presents the proposed microgrid structure, while Section 3 establishes the behavior model of EV and formulates the cost functions. Section 4 presents the algorithm of the management supervisor and shows the dynamic programming method that is utilized. The results of the management strategy are carried out in Section 5, while Section 6 concludes the work.

2. Microgrid Arrangement Description

Microgrids with PV generation have been installed in a wide power range, from single string low power rooftops ($\sim$1 kWp) up to multiple string power stations ($\sim$5 MWp). Most PV powered microgrids include multiple user loads, such as lighting, electronics, motor-based loads, and demand controlled loads, e.g., air conditioners and water heating. Recently, with the growing EV power demand, charging stations have been also added to the microgrids.

This paper considers a microgrid composed of user loads and PV arrays connected to an AC power bus (AC MG). The proposed arrangement also incorporates a few EV charging stations throughout a dedicated bus (AC EV). Both power buses are connected to the grid through a power transformer, as shown in Figure 2. For simplicity, in this paper, only the active power will be considered for the optimization purposes, and therefore, the reactive power will not be taken into account. Likewise, the distribution impedance will not be considered; however, the efficiency of all converters will be taken into account. EV users can select the charging mode on any station, so that the same station can be used for different charging modes. In Figure 2, each EV color represents a specific mode chosen by the user.

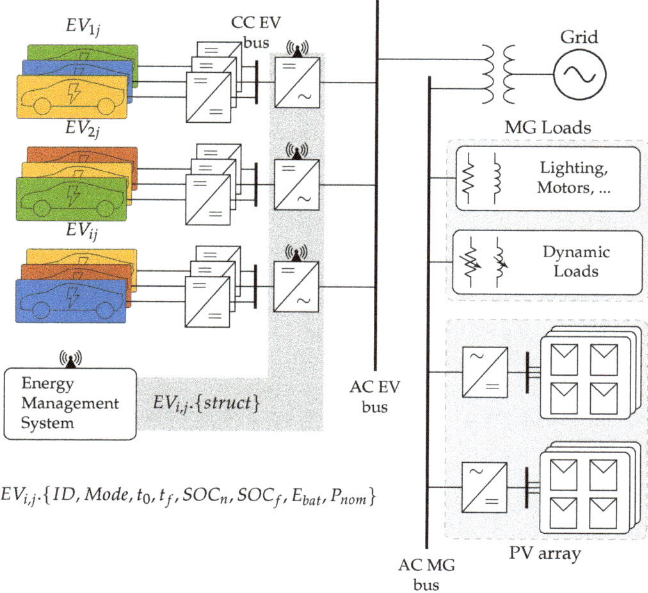

Figure 2. Physical diagram of the proposed microgrid.

Microgrid loads are usually classified into static or dynamic loads. In this paper, it will be considered a typical demand profile of a commercial establishment, and the PV generation profile will be used to represent the different weather conditions. In Figure 2, it is indicated that the EMS communicates with the converters by a communication path, and some EV information is acquired that will be used in the management.

2.1. Energy Management System

The proposed Energy Management System (EMS) block continuously collects power information from EV charging stations, the PV generation system, and microgrid demand. The EMS uses the power profile of all subsystems in order to coordinate the EV charging. To optimize the microgrid operation, EMS also collects the individual EV charging profile provided by the EV users upon arrival at the parking lot. The EV data profile consists of a data structure identified by $EV_{i,j}.\{struct\}$, which includes identification, storage system profile, and user preference parameters, wherein i is related to the charger group and j specifies the outlet of the charger. The main information in the structure is: user/vehicle identifier (ID), charging mode chosen by the user ($Mode$), connection time (t_0), expected disconnection time (t_f), current (SOC_n) and final desired SOC (SOC_f), maximum battery capacity (E_{bat}), and nominal charging power (P_{nom}).

From the microgrid diagram, in Figure 2, a simplified arrangement is proposed and brings together the EVs by charging modes (P_{EV_M}), as presented in Figure 3. The power flows in PV panels, and loads are identified by P_{PV} and P_L, respectively. In addition, the EMS receives power system measurements and dispatches the individual charging commands (P_{EV_v}) based on an optimization algorithm for all charging demands. The proposed EMS provides a set $\mathcal{M} = \{EV_U, EV_F, EV_E, EV_{V2G}\}$

of four EV charging modes: ULTRA, FAST, ECO, and V2G. The total power flows related to all vehicles in the same charging mode are given by P_U, P_F, P_E, and P_{V2G}, respectively.

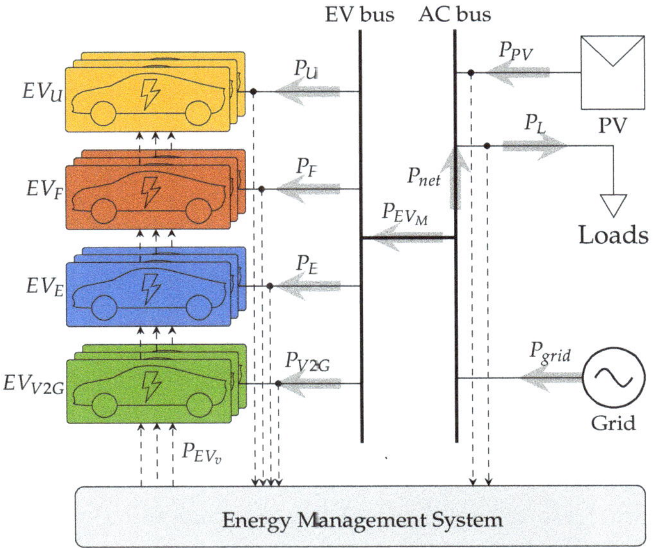

Figure 3. Simplified arrangement of the proposed microgrid.

Based on Figure 3, it can be assumed that the power balance of the proposed system arrangement is given by $P_{grid} = P_{net} + P_{EV_M}$, herein $P_{net} = P_L - P_{PV}$, and represents the resulting power between demand and generation.

The logical architecture of the EMS block is defined in Figure 4, and it consists of four functional modules: acquisition, supervision, optimization, and prediction.

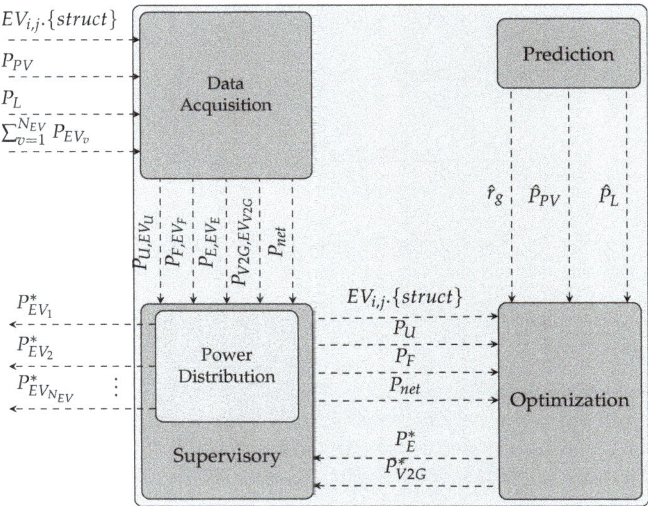

Figure 4. Proposed manager architecture.

2.2. EMS Operation

When a vehicle connects to the charger station, the EV user selects the desired charging profile, and an acquisition module reads and organizes data from EV users. Furthermore, the PV generation and load demand data are acquired continuously. The supervisor module manages the real-time tasks of the system. It receives the data from the acquisition module and verifies if it is necessary to perform an optimization operation. At each calculation step, the ULTRA and FAST power references are updated, and the supervisor module splits and sends the results to the inverter controllers. When ECO and V2G modes are selected, the supervisor also requests an optimization operation. In this case, a dynamic programming method is used to calculate the EV charging power references for all EV connection period. The optimization method uses dynamic programming in order to reach the optimal progression of charging throughout the day. In the optimization process, forecast data provided by the prediction module are used, such as energy price ($\hat{r}_g$), PV generation forecast ($\hat{P}_{PV}$), and load demand ($\hat{P}_L$).

2.3. Operational Model of EV

For simplicity, a linear battery model can be applied to obtain the SOC behavior for each vehicle, which is given by:

$$SOC_v[n+1] = SOC_v[n] + \eta_{EV}\frac{P_{EV_v}[n]\Delta t}{E_{bat_v}},\tag{1}$$

wherein E_{bat_v} is the total battery capacity to each v vehicle, η_{EV} is the charging/discharging efficiency, and SOC_v represents the resulting battery percentage one sample ahead $[n+1]$, based on charging power P_{EV_v} available on current sample n.

Fundamental restrictions are established for the SOC_v in order to adjust the EV performance and limit the control behavior P_{EV_v}. Therefore, the SOC limits are given by:

$$0 \leqslant SOC_v[n+1] \leqslant 1,\tag{2}$$

which restricts the control performance to:

$$P_{EV_v} = \begin{cases} P_{EV_v}, & \text{if } 0 \leqslant SOC_v[n+1] \leqslant 1, \\ 0, & \text{otherwise.} \end{cases}\tag{3}$$

The EV connection status is defined as:

$$EV_{conn} = \begin{cases} 1, & \text{if } EV_{i,j}.\{t_0\} \leqslant EV_v(t_m) \leqslant EV_{i,j}.\{t_f\}, \\ 1, & \text{if } SOC_v[n+1] \leqslant 1, \\ 0, & \text{otherwise.} \end{cases}\tag{4}$$

wherein $EV_{i,j}.\{t_0\}$ and $EV_{i,j}.\{t_f\}$ are arguments of the EV acquisition structure and are related to initial and final charging instants. Still, t_m indicates the current time inside the interval $[t_i, t_f]$. Initially, t_i is specified as t_0.

3. EMS Cost Factors

The cost functions are defined to achieve the EV profiles and the suitable system characteristics, e.g., peak shaving in high demand situations, desired SOC reaching, and profit from the energy exchange between the microgrid and distribution network (grid). It then discusses each part that constitutes the cost functions, so that specific features of each mode are taken into account.

3.1. State in the Final Time (ECO/V2G)

In order to establish a final value for the SOC, and thus enable a starting point for the dynamic programming method, it is proposed:

$$E_{soc} = \gamma \left(\Delta SOC_{f,EV_{E,V2G}} E_{bat,EV_{E,V2G}} \right). \tag{5}$$

Herein, E_{soc} represents the error between calculated SOC_f and the reference SOC for the last vehicle connection sample, where:

$$\Delta SOC_{f,EV_{E,V2G}} = \left(SOC_{f,EV_{E,V2G}} - SOC_{EV_{E,V2G}}(t_f) \right)^2. \tag{6}$$

The gain γ defines a weight for the final state; $\Delta SOC_{f,EV_{E,V2G}}$ represents the error between the desired SOC at final time ($SOC_{f,EV_{E,V2G}}$) and the calculated one ($SOC_{EV_{E,V2G}}(t_f)$); and $E_{bat,EV_{E,V2G}}$ is the maximum capacity of each vehicle in ECO and V2G modes.

3.2. PCC Power Balance (ECO/V2G)

In order to indicate the power balance at the Point of Common Coupling (PCC) and to verify the energy exchange between the PV source, loads, and the EV, it can established that:

$$R_{pcc} = \delta r_G(t_m) \left[P^*_{EV_{E,V2G}}(t_m) + \hat{P}_{net}(t_m) + \hat{P}_{EV_M}(t_m) \right] \Delta t. \tag{7}$$

In this case, the EV is defined as the load type profile, wherein:

$$\hat{P}_{net}(t_m) = \hat{P}_L(t_m) - \hat{P}_{PV}(t_m) \tag{8}$$

$$\hat{P}_{EV_M}(t_m) = \sum_{Mode \,\in\, \mathcal{M}^*} P_{EV_{i,j}}.\{Mode\}. \tag{9}$$

wherein $\mathcal{M}^*$ represents all vehicles in all modes, except the current vehicle.

In (7), $\delta \geqslant 1$ is the weight associated with the influence of R_{pcc} in the cost function, e.g., such that the microgrid manager could attribute an extra charge for V2G energy; $r_G(t_m)$ is the purchase energy price from the external grid for sample t_m; P^*_{EV} represents the optimal charging power. The variables $\hat{P}_L$ and $\hat{P}_{PV}$ are the one day-ahead demand and PV generation, respectively, which are obtained from the prediction module. The variable $\hat{P}_{EV_M}$ establishes the influence on the control law computation of all modes of the remaining vehicles. Therefore, the cost function takes the EV charging demand into account. The charging profile is calculated until the disconnection for all vehicles, regardless of the current instant. The term Δt is applied to equalize units of the factors of the cost function and to obtain the energy unit.

3.3. Generation Surplus (ECO/V2G)

The ECO mode is defined such the EMS commands the power stations to charge the EVs during the PV generation surplus or when the energy price is low. To accomplish this approach, a gain schedule α is added to the control response $P^*_{EV_E}$ in the PCC power balance factor. When a surplus occurs, α is lower than one, and the cost function prioritizes the charge of the EVs in both ECO and V2G modes. The surplus index is summarized as:

$$\alpha = \begin{cases} \alpha_0, & \text{if } P_{net}(t_m) + P_{EV_M}(t_m) \leqslant 0 \\ 1, & \text{otherwise.} \end{cases} \tag{10}$$

wherein $\alpha_0 < 1$ must be chosen in order to increase the charging rate when the PV generation overcomes $P_L + P_{EV_M}$.

3.4. Battery Degradation (V2G)

Since V2G mode enables the vehicle to discharge, one of the aspects that must be taken into account is the battery degradation. This is an important issue for the user's profit. This paper deals with temperature effects and Depth Of Discharge (DOD), considering the scenarios analyzed by [20,21]. Thereby, the factor associated with temperature degradation is given by:

$$R_{deg} = \frac{r_D E_{bat_v}}{L(t_m)} \Delta t \tag{11}$$

Herein, r_D is the Lithium battery price per kWh (in this paper, $r_D = 742.90$ \$/kWh), and L is the battery lifetime in seconds [3].

In Equation (11), the battery lifetime $L(t_m)$ describes the charging power influence on temperature, as:

$$L(t_m) = a\,e^{\dfrac{b}{T_{amb} + 1000R_{th}P^*_{EV_{V2G}}(t_m)}} \tag{12}$$

where a and b are constants obtained from curves that relate cycles/charging time versus 50% DOD, available in [20,21]. T_{amb} is the ambient temperature, and R_{th} is the thermal resistance. From this analysis, it is demonstrated that the higher the temperature applied on the battery, the lower is L.

The factor related to DOD degradation is considered only when discharging occurs, and it is given by:

$$E_{deg,soc} = \varepsilon\left(\Delta SOC_{dod}E_{bat_{V2G}}\right), \tag{13}$$

wherein:

$$\Delta SOC_{dod} = SOC_{possible} - SOC_{EV_{V2G}}(t_m), \tag{14}$$

The gain ε includes a relative DOD weight; ΔSOC_{dod} is the difference between possible values of SOC tested in optimization ($SOC_{possible}$) and the calculated SOC ($SOC_{EV_{V2G}}(t_m)$) according to the control law applied from the possible values of power ($Pev_{possible}$). The $SOC_{possible}$ represents all quantized levels in the SOC state variable, while $Pev_{possible}$ represents the quantized levels in the control law. Both are tested verifying all possibilities in a defined quantized space to achieve the minimum control law that results in the optimal state trajectory. The main idea of this factor is to balance the battery discharging, while minimizes the SOC difference.

3.5. Power Variation Limitation

The cost function of V2G mode also includes the charging power variation limitation, in order to avoid sudden changes in the control law. The limitation normalized by the maximum charger power is given by:

$$E_{0,\Delta P_{EV}} = \beta_0\left(\frac{P^*_{EV_{V2G}}(t_m)}{P_{nom}}\right)^2 \tag{15}$$

Herein, the gain β_0 establishes the maximum variation, and P_{nom} is the maximum charging power.

3.6. Energy Sale to the Grid

The possibility of supplying electric energy for the system through EV storage is one of the main characteristics of the proposed microgrid manager. Therefore, the cost factor of V2G mode to include the profit from the sale of energy can be established as:

$$R_{V2G} = r_{EV}(t_m) \frac{P^*_{EV_{V2G}}(t_m)}{\eta_{EV}} \Delta t. \tag{16}$$

The variable r_{EV} represents the sale price of energy, and it is defined as 0.55 \$/kWh, taking the purchase price of energy into account; and η_{EV} is the efficiency of converters. Efficiency must be exceeded for the V2G energy sale to become viable.

4. EV Charging Modes Definition

From ULTRA up to V2G charging modes, the optimization functions have a dependence on different sets of inputs. The modes' dependence is given by:

$$\begin{aligned}
\text{ULTRA} &= f(P_{net}), \\
\text{FAST} &= f(P_{net}, P_U), \\
\text{ECO} &= f(P_{net}, P_U, P_F, \$, \hat{P}_{net}), \\
\text{V2G} &= f(P_{net}, P_U, P_F, P_E, P_{V2G}, \$, \hat{P}_{net}).
\end{aligned} \tag{17}$$

The charging modes are defined to have a specific behavior and operation specifications according to the user selection. The main role and characteristics of each charging mode are shown below:

- The main stimulus of ULTRA mode is for vehicles that stay parked a short time and users that want high level priority. In this case, the nominal power charging (P_{nom}) is applied if the microgrid demand is not exceeded considering the PV/loads' power measurement (P_{net}). No prediction data are used for this mode;
- FAST mode is designed for users who want to pay less than ULTRA users for charging, but they will have reduced charging priority compared with the ULTRA ones. Analogous to the previous mode, the maximum permissible power charger is supplied, but in this case, it also takes the ULTRA demand into account. Again, no prediction data are used for this mode;
- ECO mode is formulated for users who want to pay the minimum cost to get some charging. However, these users will be considered the lowest priority. From the user specified SOC and leaving time and with the PV generation and load demand forecast, the manager optimizes the charging such that the minimum \$/kWh is applied. In this case, the final SOC is not guaranteed since it also depends on the energy cost and charging power of ULTRA and FAST vehicles, or if the user leaves before the specified departure hour.
- V2G mode may discharge the vehicle to provide profit to the user while supplying the microgrid. Although it has high flexibility, it has a complex control considering a large number of variables. Analogous to ECO mode, it is handled with an optimization method to comply with all specifications, taking the demand of all modes into account, as well as the predicted data. V2G mode can also supply power to the other vehicles when the demand of the microgrid is not sufficient to attend to ULTRA/FAST charging requests.

It is noticed that the proposed charging modes are dependent on each other, since a priority order (ULTRA > FAST > ECO) is specified to manage the different charging modes. Therefore, during the periods when the microgrid reaches its demand limitation, the charging power will be maintained on the higher priority modes while it is reduced on other modes.

4.1. ULTRA Mode

In this mode, the charging power (P_{U,EV_U}) is set to be the vehicle nominal power (P_{nom,EV_U}), whenever possible. If the microgrid reaches its contracted demand at the same t_m, the power is proportionally reduced to each vehicle in this mode, using the index $\overline{f}_U(t_m) < 1$ determined by:

$$\overline{f}_U(t_m) = \frac{P_{lim} - P_{net}(t_m)}{\sum P_{nom,EV_U}(t_m)} \tag{18}$$

wherein P_{lim} is the demand contracted by the microgrid and P_{nom,EV_U} is the nominal power charging of each vehicle in this mode. If the microgrid has already reached the contracted demand before the charging power calculation, P_U is set to zero, and all vehicles in this mode are idle. Briefly, the charging power of ULTRA mode is proposed to be:

$$P_{U,EV_U} = \begin{cases} P_{nom}, & \text{if } P_{net}(t_m) + P_U(t_m) \leqslant P_{lim} \\ \overline{P}_U(t_m), & \text{otherwise,} \end{cases} \tag{19}$$

wherein $\overline{P}_U(t_m) = \overline{f}_U(t_m) P_{nom}$.

4.2. FAST Mode

This mode follows the same strategy applied in ULTRA mode, providing the maximum charging power, taking into account that the previous mode is already achieved. Whenever a microgrid power limitation occurs, FAST mode power dispatch is limited by the index $\overline{f}_F(t_m) < 1$, given by:

$$\overline{f}_F(t_m) = \frac{P_{lim} - P_{net}(t_m) - P_U(t_m)}{\sum P_{nom,EV_F}(t_m)}, \tag{20}$$

wherein $P_U(t_m)$ indicates the sum of all vehicles in ULTRA mode. Like the previous mode, the limitation strategy is applied to this case; if the microgrid demand has already reached its limit, then P_F is set to zero. Thereby, the resultant charging power of the FAST mode may be written as:

$$P_{F,EV_F} = \begin{cases} P_{nom}, & \text{if } P_{net}(t_m) + P_U(t_m) + P_F(t_m) \leqslant P_{lim} \\ \overline{P}_F(t_m), & \text{otherwise,} \end{cases} \tag{21}$$

Herein, $\overline{P}_F(t_m) = \overline{f}_F(t_m) P_{nom}$.

4.3. ECO Mode

For the vehicles in ECO mode, the manager decides to charge them preferably when a PV surplus occurs or at periods of lower demand, therefore bringing savings for the user and benefits to the grid. From factors (Section 3) related to this mode and using the optimization technique known as dynamic programming [22], the cost function is formulated as:

$$J_E = E_{soc}(t_f) + \sum_{t_m=t_0}^{t_f} R_{pcc}, \tag{22}$$

where R_{pcc} is related to the factor presented in (7). The dispatched power (P^*_{E,EV_E}) of Equation (7) is the one that causes the minimum cost J_E among all possible power sets ($Pev_{possible}$), and it results in the optimal SOC trajectory. The definition of the optimal power (P^*_{E,EV_E}) is then given by:

$$P^*_{E,EV_E} \leftarrow J_E(P^*_{E,EV_E}) \leqslant J_E(Pev_{possible}). \tag{23}$$

In a demand limitation case, the index of the charging power is calculated from:

$$\overline{f}_E(t_m) = \frac{P_{lim} - P_{net}(t_m) - P_U(t_m) - P_F(t_m)}{\sum P_{nom,EV_E}(t_m)} \tag{24}$$

Herein, P_U and P_F are the profiles already determined in (19) and (21). Furthermore, like ULTRA and FAST modes, if the microgrid demand has reached the demand limitation, then all vehicles in this mode become idle, due to P_E being set to zero. The resultant charging power for the ECO mode vehicles is specified by:

$$P_{E,EV_E} = \begin{cases} P^*_{E,EV_E}, & \text{if } P_{net}(t_m) + P_U(t_m) + P_F(t_m) + P_E(t_m) \leqslant P_{lim} \\ \overline{P}_E(t_m), & \text{otherwise,} \end{cases} \tag{25}$$

wherein $\overline{P}_E(t_m) = \overline{f}_E(t_m) P^*_{E,EV_E}(t_m)$.

4.4. V2G Mode

Similar to ECO mode, the optimization technique and the factors previously described are used to formulate the cost function. Nevertheless, this mode is divided in two regions: $\mathcal{C}$, which represents $Pev_{possible} \geqslant 0$ (charging); and $\mathcal{D}$, which represents $Pev_{possible} < 0$ (discharging). The main idea is to deal with a specific design for each different region.

For the charging region $\mathcal{C}$, the cost function is given by:

$$J_{V2G+} = E_{soc}(t_f) + \sum_{t_m=t_i}^{t_f} \left[R_{pcc}(t_m) + R_{V2G}(t_m) + E_{0,\Delta P_{EV}}(t_m) + R_{deg}(t_m) \right], \tag{26}$$

and for region $\mathcal{D}$, it is written as:

$$J_{V2G-} = E_{soc}(t_f) + \sum_{t_m=t_0}^{t_f} \left[R_{pcc}(t_m) + R_{V2G}(t_m) + E_{0,\Delta P_{EV}}(t_m) + R_{deg}(t_m) + E_{deg,soc}(t_m) \right]. \tag{27}$$

The expressions (26) and (27) differ by the term $E_{deg,soc}(t_m)$ in the discharging region. Naturally, $J_{V2G+} > J_{V2G-}$ due to the operating charging power in the region $\mathcal{D}$ being negative, though with the addition of $E_{deg,soc}(t_m)$, the energy sale must overcome the degradation term to be feasible.

The dispatched power $P^*_{V2G,EV_{V2G}}$ is part of the set that comprises all possible quantized values ($Pev_{possible}$) and implies the total minimum cost described by:

$$J_{V2G} = J_{V2G+} + J_{V2G-}. \tag{28}$$

The optimal power definition $P^*_{V2G,EV_{V2G}}$ is given by:

$$P^*_{V2G,EV_{V2G}} \leftarrow J_{V2G}(P^*_{V2G,EV_{V2G}}) \leqslant J_{V2G}(Pev_{possible}). \tag{29}$$

In the case of the demand limitation related to vehicles in ULTRA and FAST modes and if the SOC of the EV is over 40%, the index of discharging power follows the same expression of ECO mode (see (24)). However, in this case, the V2G vehicles discharge and supply the other modes and/or the load demand. Therefore, the resulting charging power for the V2G mode is defined by:

$$P_{V2G,EV_{V2G}} = \begin{cases} P^*_{V2G,EV_{V2G}}, & \text{if } P_{net}(t_m) + P_U(t_m) + P_F(t_m) \leqslant P_{lim} \\ \overline{P}_{V2G}(t_m), & \text{otherwise,} \end{cases} \tag{30}$$

Herein, $\overline{P}_{V2G}(t_m) = \overline{f}_{V2G}(t_m) P_{nom}$.

For this situation, the current charging power of ULTRA and FAST modes is recalculated considering the P_{V2G} charging power. In a scenario where the microgrid demand has reached the contracted demand prior to calculating the charging power, then the power dispatched to the vehicles in V2G mode is $\overline{P}_{V2G}$. V2G mode vehicles may increase the virtually value of the contracted demand from the microgrid because they store energy. Thus, when there are no vehicles in V2G mode, the limitation imposed on ULTRA (see Equation (18)) and FAST (see Equation (20)) modes is performed. Otherwise, if the microgrid demand is above the limit and there is energy stored in the vehicles in V2G, the charging power of ULTRA and FAST modes will be set in the V2G algorithm.

4.5. Supervisory and Optimization Algorithms

The management strategy for the EV/PV microgrid (Figure 4) is represented by the flowchart in Figure 5, wherein the acquisition and supervision modules are highlighted.

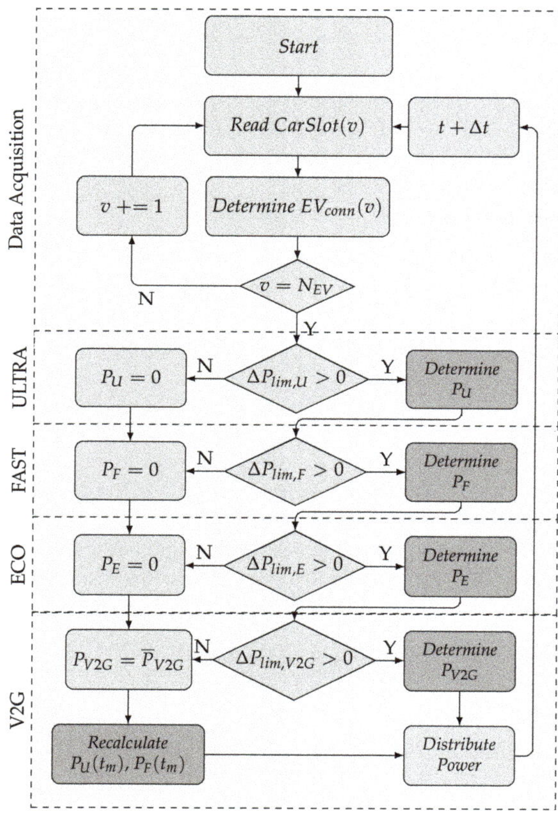

Figure 5. Management algorithm of the microgrid.

The data are acquired by EMS at each time interval, and the power P_{EV_v} for each vehicle v that is connected to the charging stations is determined. In the first step, the measurement variables are acquired, covering all parking lots ($CarSlot(v)$). When there is no vehicle connected, the algorithm follows right through the end and distributes the PV available power to loads of the microgrid. Without the EV connected, the EMS may request power from the grid or deliver PV surplus to the grid, as necessary.

When a vehicle is connected, then N_{EV} is updated, and EMS determines the connection vector EV_{conn} based on Equation (4). At the end of the acquisition process, the car data are saved using the format $EV.\{struct\}$, as defined previously.

After the acquisition process, the supervisory stage is initiated, which is responsible for defining the actions to be taken following the priority order set. This module analyzes the contracted demand limit and takes corrective actions to keep the demand at the appropriate level. The variables $\Delta P_{lim,U}$, $\Delta P_{lim,F}$, and $\Delta P_{lim,E}$ are related to the difference between the contracted demand limit (P_{lim}) and the cumulative sum of each mode. If there are vehicles in ECO or V2G modes, the optimization module is requested.

Both functions for P_{U,EV_U} and P_{F,EV_F} computation are presented in Algorithm 1, which illustrates the processing logic of the ULTRA and FAST modes.

Algorithm 1: ULTRA and FAST modes.

---

1   **if** $Mode == ULTRA$ **then**
2      $\mathcal{P} \leftarrow P_{U,EV_U}$;
3   **end**
4   **if** $Mode == FAST$ **then**
5      $\mathcal{P} \leftarrow P_{F,EV_F}$;
6   **end**
7   $SOC_{calc}[1] \leftarrow SOC(t_m)$;
8   **for** $n = 1 : (t_f - t_m) - 1$ **do**
9   Calculate $\mathcal{P}$;
10   Calculate $SOC_{calc}[n+1] \leftarrow SOC_{calc}(\mathcal{P}, SOC_{calc}[n])$;
11   **if** $SOC_{calc}[n+1] \geqslant SOC_f$ **then**
12         $\mathcal{P} \leftarrow 0$;
13   **end**
14   **end**
15   **if** $Mode==ULTRA$ **then**
16   Calculate $P_U \leftarrow \sum_{EV_U=1}^{N_U} \mathcal{P}$;
17   **if** $N_{V2G} == 0$ **then**
18   **if** $P_{net}[t_m] + P_U[t_m] \geqslant P_{lim}$ **then**
19   Calculate $\overline{P}_U(t_m)$;
20            $P_{U,N_U}[t_m] \leftarrow \overline{P}_U(t_m)$;
21   **end**
22   **end**
23   **end**
24   **if** $Mode==FAST$ **then**
25   Calculate $P_F \leftarrow \sum_{EV_F=1}^{N_F} \mathcal{P}$;
26   **if** $N_{V2G} == 0$ **then**
27   **if** $P_{net}[t_m] + P_U[t_m] + P_F[t_m] \geqslant P_{lim}$ **then**
28   Calculate $\overline{P}_F(t_m)$;
29            $P_{F,N_F}[t_m] \leftarrow \overline{P}_F(t_m)$;
30   **end**
31   **end**
32   **end**

First, the value of the current SOC is saved, and SOC_{calc} is calculated; SOC_{calc} has a variable size depending on each EV connection time. Then, the size of $\mathcal{P}$ from $(t_f - t_m)$, as well as SOC_{calc} is established, defining the charging power profile. Assuming that t_m represents the interval $[t_i, t_f]$ and that t_i is an updated value in each step, it is noted that $\mathcal{P}$ has also a varying size, which decreases each time step. The charging power profile $\mathcal{P}$ is recalculated in each time step such that its data may be used as forecast demand in the optimization method.

After the definition stage of the charging power profile, $\mathcal{P}$ is calculated according to (19) or (21), based on the specific mode. The SOC value is updated from (1), and the total demand of each mode P_U and P_F is computed. During demand limitation, the variables $\overline{P}_U$ and $\overline{P}_F$ represent the reduced charging power dispatched to the vehicles, according to the indexes $\overline{f}_U$ and $\overline{f}_F$ indicated in (19) or (21), respectively.

When ECO and V2G modes are requested, Algorithm 2 is used to perform the optimization of the ECO and V2G cost functions.

Algorithm 2: ECO and V2G modes.

---

1   **if** $Mode == ECO$ **then**
2     $\mathcal{J} \leftarrow J_{E,EV_E}$;
3     $\mathcal{P} \leftarrow P_{E,EV_E}$;
4   **end**
5   **if** $Mode == V2G$ **then**
6     $\mathcal{J} \leftarrow J_{V2G,EV_{V2G}}$;
7     $\mathcal{P} \leftarrow P_{V2G,EV_{V2G}}$;
8   **end**
9   **for** $s = 1 : SOC_{values}$ **do**
10   **for** $k = 1 : Pev_{values}$ **do**
11       $SOC_{next} \leftarrow SOC_{possible} + \eta_{EV} \dfrac{Pev_{possible}\Delta t}{E_{bat_{E,V2G}}}$;
12   **if** $(SOC_{next} \leqslant 1) \wedge (SOC_{next} \geqslant 0)$ **then**
13   Calculate $\mathcal{J}(t_f)$;
14         $Pev_{cost}[t_f - t_m, s] \leftarrow \mathcal{J}(t_f)$;
15         $Pev_{control}[t_f - t_m, s] \leftarrow Pev_{possible}[k]$;
16   **end**
17   **end**
18   **end**
19   **for** $n = (t_f - t_m) - 1 : -1 : 2$ **do**
20   **for** $s = 1 : SOC_{values}$ **do**
21   **for** $k = 1 : Pev_{values}$ **do**
22   Calculate SOC_{next};
23   Calculate $\mathcal{I}(Pev_{cost}, Pev_{control}, SOC_{possible})$;
24   Calculate $\mathcal{J}(n, \mathcal{I})$;
25   **if** $(SOC_{next} \geqslant 40\%) \vee (Pev_{possible} \geqslant 0)$ **then**
26           $Pev_{cost}[n, s] \leftarrow \mathcal{J}(n, \mathcal{I})$;
27           $Pev_{control}[n - 1, s] \leftarrow Pev_{possible}[k]$;
28   **end**
29   **end**
30   **end**
31   **end**
32   $SOC_{calc}[1] \leftarrow SOC(t_m)$;
33   **for** $n = 1 : (t_f - t_m) - 1$ **do**
34   Calculate $\mathcal{I}(Pev_{control}, SOC_{possible})$;
35   Calculate $\mathcal{P}(n, \mathcal{I})$;
36   Calculate $SOC_{calc}[n + 1] \leftarrow SOC_{calc}(\mathcal{P}, SOC_{calc}[n])$;
37   **end**
38   **if** $Mode == ECO$ **then**
39   Calculate $P_E \leftarrow \sum_{EV_E=1}^{N_E} \mathcal{P}$;
40   **if** $P_{net}[t_m] + P_U[t_m] + P_F[t_m] + P_E[t_m] \geqslant P_{lim}$ **then**
41   Calculate $\overline{P}_E(t_m)$;
42       $P_{E,N_E}[t_m] \leftarrow \overline{P}_E(t_m)$;
43   **end**
44   **end**
45   **if** $Mode == V2G$ **then**
46   Calculate $P_{V2G} \leftarrow \sum_{EV_{V2G}=1}^{N_{V2G}} \mathcal{P}$;
47   **if** $P_{net}[t_m] + P_U[t_m] + P_F[t_m] \geqslant P_{lim}$ **then**
48   Calculate $\overline{P}_{V2G}(t_m)$;
49       $P_{V2G,N_{V2G}}[t_m] \leftarrow \overline{P}_{V2G}(t_m)$;
50   Recalculate $P_{U,N_U}[t_m]$;
51   Recalculate $P_{F,N_F}[t_m]$;
52   **end**
53   **end**

The optimization related variables are quantized considering the value range and the level number of SOC (SOC_{values}) and charging power (Pev_{values}). The higher is the number of quantization levels, the more accurate will be the control, but more processing will be required. The range of allowed values is represented by $SOC_{possible}$ and $Pev_{possible}$.

In Algorithm 2, initially, all operation possibilities are examined for the last sample time (t_f) of the EV charging horizon. The costs of the optimal powers ($J(P^*_{EV}(t_f))$) are saved in Pev_{cost} for each state $SOC_{possible}$. The power that implies the lower cost of $\mathcal{J}$ is saved in matrix $Pev_{control}$, corresponding to $P^*_{EV}(t_f)$. Then, the regression analysis is performed by checking each value of $SOC_{possible}$ and $Pev_{possible}$ for all sample times in the interval $[t_m, t_f[$. From SOC_{calc} calculation, values not quantized in $SOC_{possible}$ are usually obtained; therefore, the cost is linearly interpolated ($\mathcal{I}$) according to the obtained SOC value. The control law is also interpolated. After, the cost matrix Pev_{cost} is assembled based on lower calculated cost ($J(P^*_{EV}(t_m - t_f))$). The optimal power matrix $Pev_{control}$ receives the control law, which provides the lower cost ($P^*_{EV}(t_m - t_f)$).

The progressive stage (Line 32) follows the same strategy adopted for ULTRA and FAST; however, the matrix $Pev_{control}$ calculated in the previous stage is used, which contains the optimal power profile for the determined initial SOC. This optimization process is repeated throughout all the EV connection period, including the charging profiles to the P_E and P_{V2G} vectors.

The demand limitation is verified for both ECO and V2G modes, and when V2G mode is requested, the charging dispatch for ULTRA ($\overline{P}_U$) and FAST ($\overline{P}_F$) modes is recalculated.

4.6. Prediction Module

To verify the EMS operation, the same values of PV generation and demand are utilized, however considering a 30 minute sample time decimation to add some issues for the management. These data are acquired to represent a typical scenario of a microgrid profile and to verify the EMS response during the PV and load variations. Therefore, the variables $\hat{P}_{PV}$ and $\hat{P}_L$ have the same behavior of P_{PV} and P_L except for the different sample time between them.

The EMS is considered to be able to obtain the data of the energy price ($\hat{r}_g$) from an Internet connection; thereby, the system may be subject to market price changes.

In the emulation environment, an additional Python algorithm was developed, such that it can collected forecast data of PV generation using an Internet-based communication.

5. EMS Testing Results

In order to verify the proposed management strategy, at first, the manager architecture (see Figure 4) was entirely implemented using MATLAB® software with a conceptual microgrid (see Figure 2). Additionally, the EMS algorithms were also implemented using the Python language and executed in a Raspberry Pi 3 platform. A MicroLabBox dSPACE was used to verify the real-time EMS behavior on a microgrid emulation platform, and its schematic is presented in Figure 6.

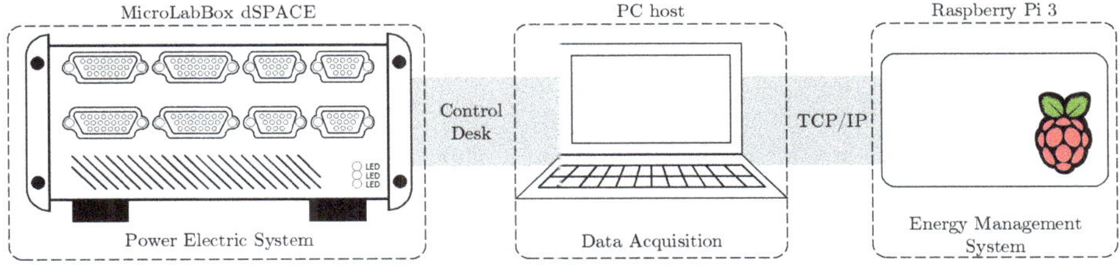

Figure 6. Platform schematic of system emulation with MicroLabBox dSPACE, a Raspberry Pi 3, and a PC host.

The power electric system was emulated within the dSPACE, and all required data were made available by the software dSPACE/ControlDesk. Since the data must be managed by the Raspberry, a TCP/IP protocol was chosen for the communication path. The energy management was computed by the Raspberry, then all reference signals were sent back to the power system emulation platform, through the PC host. To verify the EMS operation, the emulation results obtained by the platform, illustrated in Figure 6, were compared and matched to the simulation results that are presented in this section.

Figure 7 presents the measured data along with the energy price curve provided by [23], based on the commercial mode fee. The predicted data were determined using the measured data for a week before with 30 minute sample time decimation. Since the main scope of this paper is on the optimization and management logic, the EMS evaluation considered a constant temperature value throughout the day T_{amb} in order to highlight the effect of main variables P_{PV}, P_L, r_G, and EV charging operation.

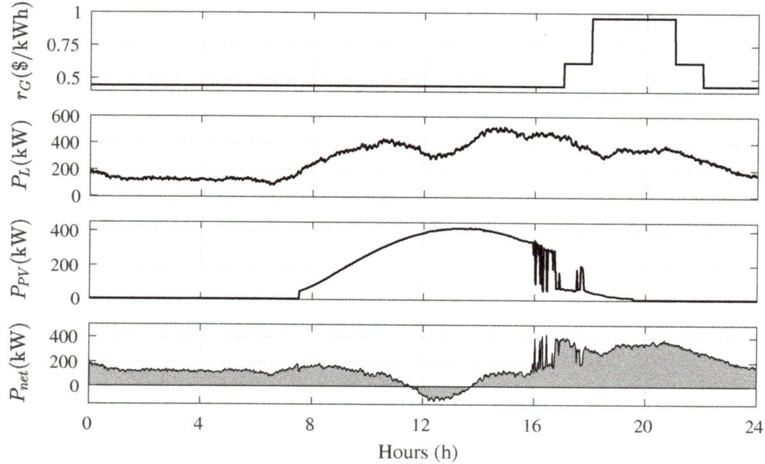

Figure 7. Microgrid profile used in EMS analysis: price of the grid energy (r_G), load demand (P_L), PV generation (P_{PV}), and the net PCC power (P_{net}).

Ten EV slots were considered in the facility parking lot, with distributed modes and parking periods. This arrangement was defined in order for it to be possible to identify the manager performance against connection variability and operation modes. Table 1 presents the EMS parameters used in the analysis, while the EV charger was considered a commercial model Terra 54, developed by ABB company, which provided 50 kW of nominal power [24]. It can be seen that $Pev_{possible}$ had two different ranges, and the total number of values (quantization) was determined by Pev_{values}, as for SOC range, which was defined by SOC_{values}.

Table 1. Simulation and real-time emulation parameters.

Energy Management System Parameters			
a	3.145×10^6	η	94%
b	-6.951	P_{nom}	50 kW
T_{amb}	25 °C	P_{lim}	450 kW
R_{th}	0.002 °C/W	N_{EV}	10
γ	1	E_{bat}	50 kWh
δ	1	Δt	1/60
ϵ	1	H	1440
α_0	0.5	SOC_{values}	10
β_0	1	Pev_{values}	11
r_D	742.90 \$/kWh	$Pev_{possible}(ECO)$	$0 \rightarrow 0.5P_{nom}$
r_{EV}	0.55 \$/kWh	$Pev_{possible}(V2G)$	$-P_{nom} \rightarrow P_{nom}$

In order to validate the algorithms developed for management, microgrid evaluations were performed using the the proposed EMS, the Terra 54 charger, a 420 kWp PV distributed generation, and a 450 kW load demand. Figure 8 shows the microgrid power flows, emphasizing the V2G mode, which is separated by two regions: charging (V2G$^+$) and discharging (V2G$^-$). The required of the microgrid from the external grid (P_{grid}) is highlighted in red.

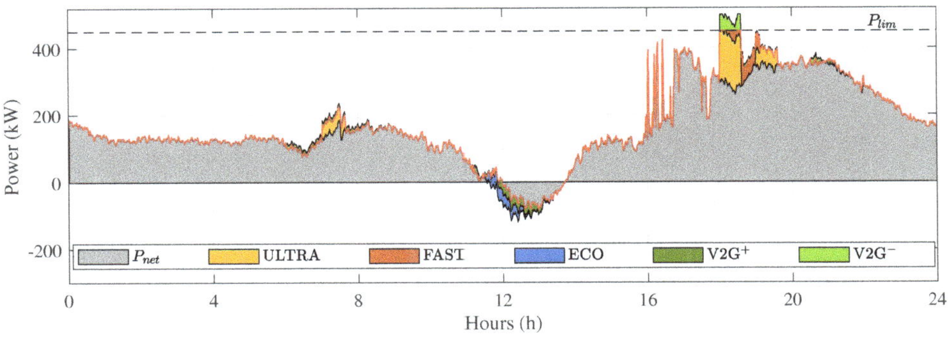

Figure 8. Power flow of charging modes and the liquid power delivered to the PCC, highlighting the external grid power (P_{grid}) with the red line.

When V2G$^-$ was present, part of the microgrid demand was supplied by the vehicles in V2G mode. During the surplus period, it can be noticed that the energy was used more efficiently, being applied to charge those vehicles set on ECO and V2G modes instead of being exported to the external grid. Hence, it is important to note that vehicles set on V2G mode only charged when the energy price was appropriate. Moreover, between 6 p.m. to 8 p.m., the V2G stored energy was discharged, generating economic benefits for the microgrid owner such as the energy purchase in moments of peak demand, reducing the facility operation costs. Figure 9 shows the charging mode (1, ULTRA; 2, FAST; 3, ECO; 4, V2G), power flow, and SOC for each EV in the parking lot. The gray regions represent the parking duration for each EV, while the solid line presents the instantaneous power and SOC defined/optimized by the EMS.

An analysis of each EV indicated that the ones in ULTRA mode were charged at nominal power while the vehicles in FAST mode had a slight reduction during the period of limited demand. Vehicles in ECO mode were charged at periods with PV surplus, avoiding demand peaks. In addition, EVs on V2G mode supplied energy to all other vehicles during moments of peak demand, around 6 p.m. To achieve this condition, the EMS commanded the energy surplus storage from 11:30 a.m. to 1:30 p.m. to sell it for a higher price during demand peaks. Since vehicles in V2G mode included discharging possibilities, the SOC and power flow of these vehicles are depicted individually in Figure 10.

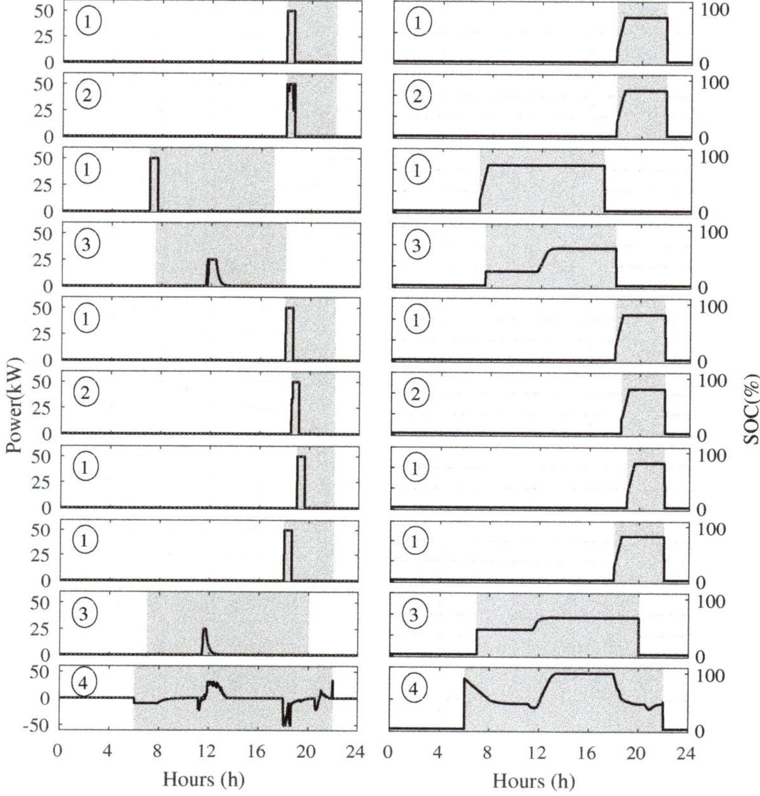

Figure 9. EV instantaneous power (**left**) and corresponding SOC (**right**), with the charging mode and the connection period. The modes are represented by: 1, ULTRA; 2, FAST; 3, ECO; and 4, V2G.

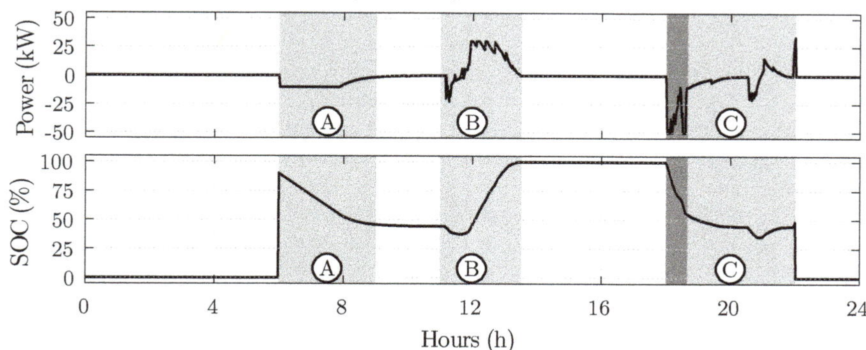

Figure 10. Charging power profile and SOC of a vehicle in V2G mode.

Region A in the Figure 10 represents the initial EV_{V2G} discharging. The discharge was related to the high initial EV_{V2G} SOC and to the predicted charging period in Region B, where a PV surplus existed, and the energy cost was lower considering the action of α proposed in Generation Surplus subsection. In Region C, the price of the grid energy had its highest value, so it was convenient that EV_{V2G} vehicles assisted in power supply for the microgrid. The charging power spikes close to the disconnection instant were related to the factor state in the final time. The dark part in Region C was associated with the demand limitation, since in this condition, the EV_{V2G} was commanded up to its nominal power in order to supply the microgrid overdemand power.

In the presented evaluation, there was a charging limitation only for FAST mode, confirming the priority employed by ULTRA mode. ECO mode also met the priority set and achieved an optimal charging profile without exceeding demand limits. V2G mode fulfilled the preferences defined by the user and the facility, and a more efficient use of the energy from the microgrid was obtained. The overall influence of the EMS for EV parking lot can be seen in Figure 11, where the P_{net} and P_{grid} flows are shown individually.

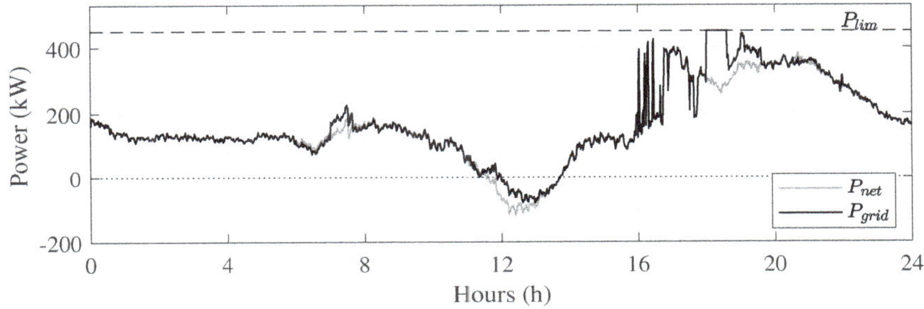

Figure 11. Resulting power between demand and generation and the power exchange with the external grid considering the EVs.

It can be seen that EV charging increased the microgrid demand at night, but the effect of valley filling was obtained throughout the afternoon. The proposed EMS behaved appropriately to supply the demand peaks, both limiting the charging of the EV_F, as well as providing the valley filling from PV surplus using EV_{V2G}.

A comparison of numerical analysis is presented in Table 2, wherein four scenarios are shown: 1, the microgrid without the EV parking lot; 2, with the EV parking lot, but no EMS; 3, with EV with a conventional EMS, with only ULTRA and FAST modes; and 4, with the EV parking lot with the proposed EMS. All scenarios that utilized EV slots took the configuration presented in Figure 9 into account, except for some changes among the scenarios. The second scenario considered EV only in ULTRA mode, without any EMS strategy. The third scenario considered a conventional EMS where only ULTRA and FAST modes were applied. ECO mode EVs in Figure 9 were considered in FAST mode, and the vehicle in V2G mode was assumed in ULTRA mode. The last scenario applied the proposed EMS entirely.

Table 2. Comparison for the operation of the parking lot with different EMS scenarios.

Scenarios	Time Above the Contracted Demand (%)	Energy Cost for Grid Purchase ($)	Relative Cost (%)
1	0%	2361.85	——
2	2.5%	2856.23	+20.93%
3	0%	2569.66	+8.8%
4	0%	2537.39	+7.43%

It can be seen that the system without an EV parking lot had the minimum cost with grid purchase; however, the increase of the EV market will make this arrangement more common. Then, if the system has no EMS to manage the charge/discharge, the price spent may rise, as can be noticed in the second scenario, which presented a cost increase of 20.93%. This cost was considered a penalty for exceeding the demand limit [25] along 2.5% of the day above its demand limit. This situation was improved with the conventional EMS, which reduced back to 0% the exceeding the energy of the microgrid, and decreased to 8.8% the cost of grid energy purchase. The influence of ECO and V2G modes can be seen in the fourth scenario, which reduced the energy cost to 7.43%, maintaining the system demand within the limits.

6. Conclusions

This paper proposed a new approach for energy management systems for microgrids with loads, PV units, and EV parking lots. The management system provided four charging modes that could be chosen by the users of the parking lot, and the EV factors that must be considered in charging modes were formulated.

The proposed EMS was evaluated using a 450 kW typical microgrid. As designed, the ULTRA mode charged the vehicles with nominal power within the expected time. In the FAST mode, the charging level and the priority level were achieved, since the EMS applied a charging limitation when demand exceeded the microgrid capacity. ECO mode was designed to provide the minimum charging cost to the user; thereby, over the analysis, it was verified that the vehicle charging occurred along the PV surplus, and no extra demand was requested from the microgrid in high demand periods. The V2G mode was developed to provide several benefits to the microgrid and also to attend to the user preferences. The V2G vehicles were discharged in high demand periods to support the microgrid loads and/or the other modes, and the same ones were charged in the PV surplus period. This performance occurred due to the relation between the price of V2G and the grid energy, taking the battery degradation into account to determine the charging power profile. The optimization algorithms were executed in convenient time in a Raspberry Pi 3 platform and were tested with an dSPACE emulation environment. All EV user preferences were met, as well as a relief in the valley of the microgrid demand, providing economic benefits to the microgrid agent.

Author Contributions: The authors contributed equally in the research: Conceptualization: E.G.C. and G.A.S.; Formal analysis: G.A.S., R.C. and C.M.d.O.S.; Investigation: E.G.C., G.A.S. and J.P.d.C.; Project administration: J.P.d.C.; Resources, C.M.d.O.S.; Supervision: J.P.d.C.; Validation: G.A.S.; Writing–original draft: E.G.C. and G.A.S.; Writing–review and editing: E.G.C., R.C. and C.M.d.O.S. All authors have read and agreed to the published version of the manuscript.

Funding: This work was supported by the Research and Development project PD 2866-0468/2017, granted by the Brazilian Electricity Regulatory Agency (ANEEL) and Companhia Paranaense de Energia (COPEL).

Acknowledgments: The authors thank the Funding Authority for Studies and Projects (FINEP), the Coordination for the Improvement of Higher Level-or Education-Personnel (CAPES), the Brazilian National Council for Scientific and Technological Development (CNPq), the Fundação Araucária, and the Universidade Tecnológica Federal do Paraná (UTFPR) for their complementary support to this work.

Conflicts of Interest: The authors declare no conflict of interest. The funders had no role in the design of the study; in the collection, analyses, or interpretation of data; in the writing of the manuscript; nor in the decision to publish the results.

Abbreviations

The following abbreviations are used in this manuscript:

DG	Distributed Generation
ESS	Energy Storage System
EV	Electric Vehicle
EMS	Energy Management System
PV	Photovoltaic
PCC	Point of Common Coupling
V2G	Vehicle-to-Grid

MPPT	Maximum Power Point Tracking
SOC	State of Charge
MPC	Model Predictive Control
ARIMA	Autoregressive Integrated Moving Average
MILP	Mixed-Integer Linear Programming
AC	Alternative Current
DC	Direct Current
MG	Microgrid
DOD	Depth of Discharge

References

1. Luna, A.C.; Meng, L.; Diaz, N.L.; Graells, M.; Vasquez, J.C.; Guerrero, J.M. Online Energy Management Systems for Microgrids: Experimental Validation and Assessment Framework. *IEEE Trans. Power Electron.* **2018**, *33*, 2201–2215. [CrossRef]
2. Mahmud, K.; Hossain, M.J.; Town, G.E. Peak-Load Reduction by Coordinated Response of Photovoltaics, Battery Storage, and Electric Vehicles. *IEEE Access* **2018**, *6*, 29353–29365. [CrossRef]
3. Tsiropoulos I.; Tarvydas D.; Lebedeva N. *Li-Ion Batteries for Mobility and Stationary Storage Applications-Scenarios for Costs and Market Growth*; Publications Office of the European Union: Brussels, Belgium, 2018. [CrossRef]
4. Chukwu, U.C.; Mahajan, S.M. V2G Parking Lot With PV Rooftop for Capacity Enhancement of a Distribution System. *IEEE Trans. Sustain. Energy* **2014**, *5*, 119–127. [CrossRef]
5. Yan, Q.; Zhang, B.; Kezunovic, M. Optimized Operational Cost Reduction for an EV Charging Station Integrated With Battery Energy Storage and PV Generation. *IEEE Trans. Smart Grid* **2019**, *10*, 2096–2106. [CrossRef]
6. van der Meer, D.; Chandra Mouli, G.R.; Morales-España Mouli, G.; Elizondo, L.R.; Bauer, P. Energy Management System With PV Power Forecast to Optimally Charge EVs at the Workplace. *IEEE Trans. Ind. Informat.* **2018**, *14*, 311–320. [CrossRef]
7. Kikusato, H.; Mori, K.; Yoshizawa, S.; Fujimoto, Y.; Asano, H.; Hayashi, Y.; Kawashima, A.; Inagaki, S.; Suzuki, T. Electric Vehicle Charge–Discharge Management for Utilization of Photovoltaic by Coordination Between Home and Grid Energy Management Systems. *IEEE Trans. Smart Grid* **2019**, *10*, 3186–3197. [CrossRef]
8. Arias, N.B.; Hashemi, S.; Andersen, P.B.; Træholt, C.; Romero, R. Distribution System Services Provided by Electric Vehicles: Recent Status, Challenges, and Future Prospects. *IEEE Trans. Intell. Transp. Syst.* **2019**, *20*, 1–20. [CrossRef]
9. Shaukat, N.; Khan, B.; Ali, S.M.; Mehmood, C.A.; Khan, J.; Farid, U.; Majid, M.; Anwar, S.M.; Jawad, M.; Ullah, Z. A survey on electric vehicle transportation within smart grid system. *Renew. Sustain. Energy Rev.* **2018**, *81*, 1329–1349. [CrossRef]
10. European Environment Agency. *Electric Vehicles in Europe. European Environmental Agency*; Report No 20/2016; Publications Office of the European Union: Brussels, Belgium, 2016 [CrossRef]
11. Chandra Mouli, G.R.; Kefayati, M.; Baldick, R.; Bauer, P. Integrated PV Charging of EV Fleet Based on Energy Prices, V2G, and Offer of Reserves. *IEEE Trans. Smart Grid* **2019**, *10*, 1313–1325. [CrossRef]
12. Hamidi, A.; Nazarpour, D.; Golshannavaz, S. Multiobjective Scheduling of Microgrids to Harvest Higher Photovoltaic Energy. *IEEE Trans. Ind. Informat.* **2018**, *14*, 47–57. [CrossRef]
13. Melhem, F.Y.; Grunder, O.; Hammoudan, Z.; Moubayed, N. Optimization and Energy Management in Smart Home Considering Photovoltaic, Wind, and Battery Storage System With Integration of Electric Vehicles. *Can. J. Elect. Comput. Eng.* **2017**, *40*, 128–138. [CrossRef]
14. Garcia-Torres, F.; Vilaplana, D.G.; Bordons, C.; Roncero-Sanchez, P.; Ridao, M.A. Optimal Management of Microgrids with External Agents including Battery/Fuel Cell Electric Vehicles. *IEEE Trans. Smart Grid* **2018**, *10*, 4299–4308. [CrossRef]
15. Tushar, M.H.K.; Zeineddine, A.W.; Assi, C. Demand-Side Management by Regulating Charging and Discharging of the EV, ESS, and Utilizing Renewable Energy. *IEEE Trans. Ind. Informat.* **2018**, *14*, 117–126. [CrossRef]

16. Wang, J.; Bharati, G.R.; Paudyal, S.; Ceylan, O.; Bhattarai, B.P.; Myers, K.S. Coordinated Electric Vehicle Charging With Reactive Power Support to Distribution Grids. *IEEE Trans. Ind. Informat.* **2019**, *15*, 54–63. [CrossRef]

17. Latifi, M.; Khalili, A.; Rastegarnia, A.; Sanei, S. A Bayesian Real-Time Electric Vehicle Charging Strategy for Mitigating Renewable Energy Fluctuations. *IEEE Trans. Ind. Informat.* **2019**, *15*, 2555–2568. [CrossRef]

18. Zheng, Y.; Song, Y.; Hill, D.J.; Meng, K. Online Distributed MPC-Based Optimal Scheduling for EV Charging Stations in Distribution Systems. *IEEE Trans. Ind. Informat.* **2019**, *15*, 638–649. [CrossRef]

19. Akhavan-Rezai, E.; Shaaban, M.F.; El-Saadany, E.F.; Karray, F. New EMS to Incorporate Smart Parking Lots Into Demand Response. *IEEE Trans. Smart Grid* **2018**, *9*, 1376–1386. [CrossRef]

20. Hoke, A.; Brissette, A.; Maksimović, D.; Pratt, A.; Smith, K. Electric vehicle charge optimization including effects of lithium-ion battery degradation. In Proceedings of the 2011 IEEE Vehicle Power and Propulsion Conference, Chicago, IL, USA, 6–9 September 2011; pp. 1–8. [CrossRef]

21. Hoke, A.; Brissette, A.; Smith, K.; Pratt, A.; Maksimovic, D. Accounting for Lithium-Ion Battery Degradation in Electric Vehicle Charging Optimization. *IEEE J. Emerg. Sel. Topics Power Electron.* **2014**, *2*, 691–700. [CrossRef]

22. Kirk, D. Optimal Control Theory: An Introduction. In *Dover Books on Electrical Engineering Series*; Dover Publications: Mineola, NY, USA, 2004.

23. COPEL. Tarifa Branca-Copel. 2019. Available online: https://www.copel.com/hpcopel/root/nivel2.jsp?endereco=%2Fhpcopel%2Findustrial%2Fpagcopel2.nsf%2Fdocs%2FB0CA4C8DF4B62F98832581F00058CCF9 (accessed on 3 February 2020).

24. ABB. Electric Vehicle Infrastructure Terra 54 and Terra 54HV UL DC Fast Charging Station. 2019. Available online: https://search-ext.abb.com/library/Download.aspx?DocumentID=4EVC800801-LFUS&LanguageCode=en&DocumentPartId=&Action=Launch (accessed on 3 February 2020).

25. COPEL. Demanda-Demanda de Ultrapassagem. 2019. Available online: https://www.copel.com/hpcopel/root/nivel2.jsp?endereco=%2Fhpcopel%2Froot%2Fpagcopel2.nsf%2Fdocs%2FF5EAD992942579F903257EBB0042F764#DU (accessed on 3 February 2020).

Review

A Review on Communication Standards and Charging Topologies of V2G and V2H Operation Strategies

Seyfettin Vadi [1], Ramazan Bayindir [2,*], Alperen Mustafa Colak [3] and Eklas Hossain [4,*]

[1] Department of Electronics and Automation, Vocational School of Technical Sciences, Gazi University, Ankara 06500, Turkey; seyfettinvadi@gazi.edu.tr
[2] Department of Electrical and Electronics Engineering, Faculty of Technology, Gazi University, Ankara 06500, Turkey
[3] Department of Advanced Technology and Science for Sustainable Development, Graduate School of Engineering, Nagasaki University, 1-14 Bunkyo, Nagasaki 852-8521, Japan; colak@pca.cis.nagasaki-u.ac.jp
[4] Department of Electrical Engineering and Renewable Energy, Oregon Renewable Energy Center (OREC), Oregon Tech, Klamath Falls, OR 97601, USA
*   Correspondence: bayindir@gazi.edu.tr (R.B.); eklas hossain@oit.edu (E.H.); Tel.: +1-541-885-1516 (E.H.)

Received: 22 August 2019; Accepted: 26 September 2019; Published: 30 September 2019

Abstract: Electric vehicles are the latest form of technology developed to create an environmentally friendly transportation sector and act as an additional energy source to minimize the demand on the grid. This comprehensive research review presents the vehicle-to-grid (V2G) and the vehicle-to-home (V2H) technologies, along with their structures, components, power electronic topologies, communication standards, socket structure, and charging methods. In addition, the charging topologies in V2G and V2H are given in detail. This study is planned as a useful guide for future studies that can be achieved in that it compares the results obtained and analyzes the studies in the literature, finding the advantages and disadvantages of charging topologies in V2G and V2H.

Keywords: vehicle-to-grid (V2G); vehicle-to-home (V2H); bi-directional charging topologies; communication standards; battery cycle

---

1. Introduction

The vehicle-to-grid (V2G) and vehicle-to-home (V2H) technologies refer to the transfer of electricity from vehicles to the grid and the home, respectively. To learn how these technologies have come about, we have studied some of the background literature on the topic. In recent years, because of the newly emerging needs parallel to developing technology and industrialization, the increasing awareness of global energy consumption, global warming, and environmental issues has in many countries forced authorities to make decisions about greenhouse gas emissions. In addition, the constant increase in crude oil prices has led societies to look for alternative fuel types and seek ways to rescue them from oil dependence [1]. Hydrogen fuel cells are a solid option for the future of fuels. Hydrogen is being seen opportunistically by many gas and water companies in Germany for its economic potential in gas-grid networks, facilitating what is known as power-to-gas [2]. Although the hydrogen economy seems propitious, hydrogen itself causes some unresolved problems. Fuel cells are expensive and require a completely new distribution network. Moreover, hydrogen storage also causes certain difficulties, as it is an explosive and a sensitive gas that further limits the large-scale prevention due to its low density [3]. Hydrogen is not found in pure form such as with oil or coal. Thus, hydrogen is bound to another element in nature. For example, water is obtained from the chemical reaction of hydrogen with oxygen, aided by a spark of energy. For this reason, pure hydrogen must be produced in a way that requires energy, such as gasoline. Because of the high energy losses within a hydrogen economy, the synthetic energy carrier cannot compete with electric energy [4]. Thus, electric energy may be considered as an alternative to conventional fuels. There are both advantages and disadvantages of electricity when it is stored. However, electricity is the most exceptional technology for the future because its infrastructure is ready and its technology is considered to be safe and reliable technology. Moreover, constant increasing fuel prices and emerging environmental awareness have forced societies to look for alternative transport solutions [5]. In the past, electric vehicles (EV) were not a strong alternative for internal-combustion engines (ICE) because of their weak batteries. However, recently batteries and EVs have been developed that have become important alternatives for conventional vehicles. Although battery prices are high and driving ranges are lower compared to conventional vehicles, EVs offer many direct and indirect benefits for society. There are many electricity transportation projects running all over the world, and EV technology and its trade are developing rapidly [6]. For this reason, it is extremely important to act in line with this technology and to obtain further innovations and concept developments [7].

Currently, transportation infrastructure consumes 26% of the annually produced energy; almost all of these energies are from fossil fuels. A total of 11% of the CO_2 emission of the world is caused by individual use on roads. The current CO_2 concentration that exists in the atmosphere is 286 ppm, and it is increasing at a rate of 2 ppm each year. Under normal conditions, it is predicted that it might be 700 ppm in some regions by 2100 [8]. It is thus of crucial importance that radical changes be made in the usage of fossil fuels to lower CO_2 emissions in the atmosphere, otherwise the earth will soon perish. The existing manufacturers must enforce new solutions to ensure efficient fuel consumption with legal restrictions on CO_2 emissions. The most important factor that will support these efforts is the widespread use of EVs for individual use [9]. Mass shifting to EVs, although a time-staking initiative, will significantly reduce CO_2 emissions, also having numerous added benefits.

EVs store large amounts of energy in a reliable form. Classic electric vehicles may be considered a form of load that charges and uses the batteries from the electricity grid, and charging is a one-way process from the grid to the vehicle [10]. The rate of using electric vehicles increases with each passing day and constitutes an additional burden on the existing grid. In this case, there is a need for extra investment in the infrastructure, and costs are quite high. At this point, the idea of using EVs as an energy source has come to the agenda. In this way, additional investment will be avoided in the existing system [11]. Today, electric vehicles transfer energy to the grid when the vehicle will not be used for a long time or when the electricity sales price is at its highest. This technology is called V2G. Users can charge the battery when energy prices are low and provide both the current energy demand of the grid and gain financial benefits. In the literature, this process is named grid-to-vehicle (G2V) technology. If the energy of the electric vehicle battery is supplied to individual houses instead of the grid, it is called as the V2H technology [12]. A general operation mode of the G2V, V2G, and V2H technologies are given in Figure 1.

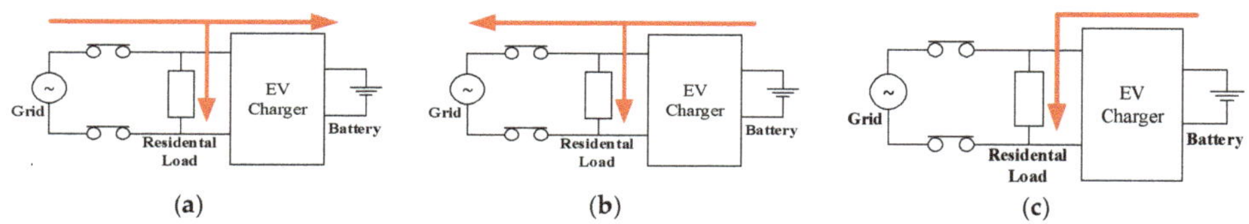

Figure 1. Operating modes of the bi-directional electric vehicle (EV) charger. (**a**) grid-to-vehicle (G2V) mode; (**b**) vehicle-to-grid (V2G) mode; (**c**) vehicle-to-home (V2H) mode [12].

Because V2G and V2H technologies are the fastest methods of meeting the increasing energy demand and can make an investment in the existing infrastructure, internationally recognized standards have been established. Thanks to these standards, a common language has been created among countries [13].

With the use of EVs together with renewable energy sources, the micro-grid structure is diversified in terms of relevant energy sources [14]. In this respect, as illustrated in Figure 2, the vertical bold line represents the grid, to which several power converters are connected, which converts the AC power of the grid into AC or DC of suitable magnitude for a device or machine connected to it, such as solar array, windmill, vehicles, uninterrupted power supplies (UPS), motor, lighting, appliances, generator, or FACTS devices. The figure shows that EVs are used as an energy source as well as a load when the user desires. This has also established a dynamic grid structure that ensures energy continuity [15].

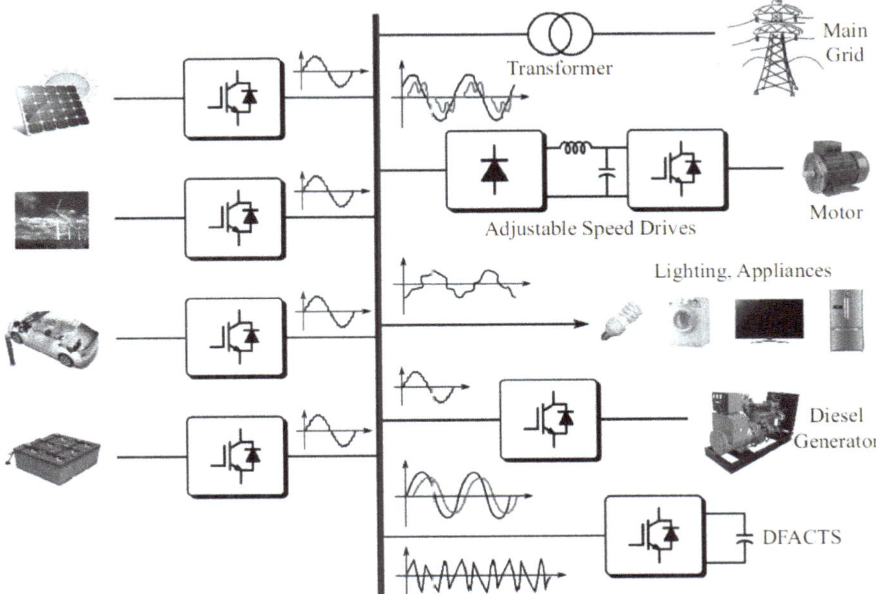

Figure 2. A general system structure for microgrid.

In this study, research in the literature and application types and standards have been investigated in order to operate EVs interactively with the grid and with houses. The grid-interactive bidirectional charging topologies are examined, and the advantages and disadvantages of these topologies are determined. Moreover, the system components are examined, and information on these components are given in detail. The rest of the sections are arranged as follows: Section 2 provides an overview of and introduction to the vehicle-to-X (V2X) technology in general, along with specific description of the V2G and V2H technologies. Section 3 outlines the many different charging topologies of the batteries of the EVs used in V2X technology. Section 4 contains an introduction to the communication standards used in V2X technologies, along with specifications of several such standards. Section 5 is composed of an elaborate discussion about the different types of AC/DC and DC/DC power converters, along with a comparative view of the types. This section is also embellished with a SWOT analysis of the V2G and V2H technologies. Next, the outcomes of this paper is discussed in Section 6. Finally, the conclusions are drawn in Section 7.

2. Vehicle to X (V2X)

Prior to learning about the V2G and V2H technologies mentioned in the title, it is important to know about their brothers, generalized by the term V2X, where X is a variable. When studies conducted in the literature are examined, it is seen that the name of the energy transfer technology depends on the target system to which the energy produced by the electric vehicle is transferred, that is, the recipient of energy [16]. As depicted in Figure 3, if an electric vehicle transfers the energy to another electric vehicle, the technology is called vehicle-to-vehicle (V2V); if the electric vehicle charges an electronic device, the technology is called vehicle-to-device (V2D); if the electric vehicle transfers the energy to the grid, the technology is named V2G; if the electric vehicle provides energy to a house, the technology is called V2H, or if the energy is transferred to a building, the technology is referred to as vehicle-to-building (V2B) [17]. In general, V2X technology provides a safe, sustainable, and competitive energy supply [18].

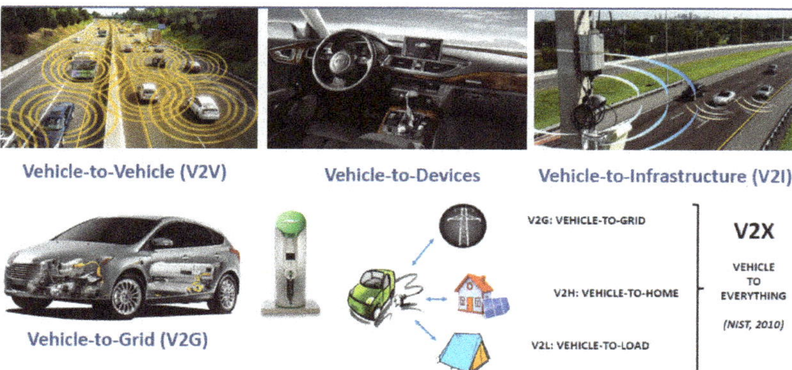

Figure 3. Different interaction modes of electric vehicles [18].

The amount of energy transferred from the vehicle to the grid depends on the number of vehicles in connection. For instance, V2H requires at least 1 or 3 vehicles, V2B requires at least 1 or 30 vehicles, and V2G requires at least 5 or 50 vehicles. The energy transfer from vehicle to the grid has to be achieved smoothly without changing the voltage, the power factor, or the frequency of the grid [19].

2.1. Vehicle to Grid (V2G)

V2G is the technology consisting of energy transfer in two directions, either from the vehicle to the grid if the energy stored in the battery is high, or from the grid to the vehicle when the energy stored in the battery is low [20]. The general block diagram of V2G structure is depicted in Figure 4. Since the energy flows to the vehicle from the grid, and from the grid to the vehicle, there occurs a bi-directional energy flow. During the energy flow, conversion operations are made with power electronics circuits to fit the type of power [21]. For the purpose of taking the voltage level of the current grid voltage to the appropriate level, firstly an AC/DC conversion system is carried out, and then the reduction is done with a descending converter. To transfer the DC energy in the battery to the grid, firstly, an additional operation is done with an increasing converter, and then the DC/AC conversion process is carried out. These operations are performed bi-directionally depending on the amount of energy in the battery. The purpose is to meet the energy demand by managing the energy in the battery or the grid [22].

The integration of plug-in hybrid vehicles (PHEV) with smart grids is also drawing significant attention of the scientific community because of its potential role in enhancing grid efficiency [23,24]. Hybrid electric vehicles usually charge their batteries through fuel, whereas electrical vehicles charge their batteries from the grid, thus having significant differences in charging circuitry and strategy [25]. However, both are of considerable interest in V2G technology. The EV is plugged into a charge station, which is connected to a power converter, followed by a transformer connected to the main grid. A control and monitoring unit is present to provide accurate input to the converter by comparing the reference signal and the outputs of the power converter (because the converter is bidirectional, both its sides can be considered as the output side). Given that electric vehicles are integrated into electric energy, they are considered as an alternative energy source for the grid. Moreover, they are also used as uninterrupted power supplies (UPS) [26]. The vehicles consisting of V2G technology are generally charged at times when electricity production is higher, or at times when the price of electricity is low at peak load times, selling the energy to the grid at peak load times with high prices or when there is energy demand. In addition, this system also provides extra energy to the grid, increasing the reliability and efficiency of the energy system [27]. The carbon emission level of V2G technology is very low, and it also works in compliance with renewable energy sources. One of the problems experienced by electric vehicles in transferring energy to the grid is the coordination with the grid operators, and another problem is the bidirectional energy and communication infrastructure [28]. Several modes of communication, such as global positioning systems [29,30] and cellular networks [31–33], are merged with the vehicles to take advantage of the vehicles in every way possible. Different intra-vehicle and inter-vehicle communication and wired or wireless protocols are established to analyze the viability, although they are still at their primary stage, requiring more contribution for greater enrichment [34].

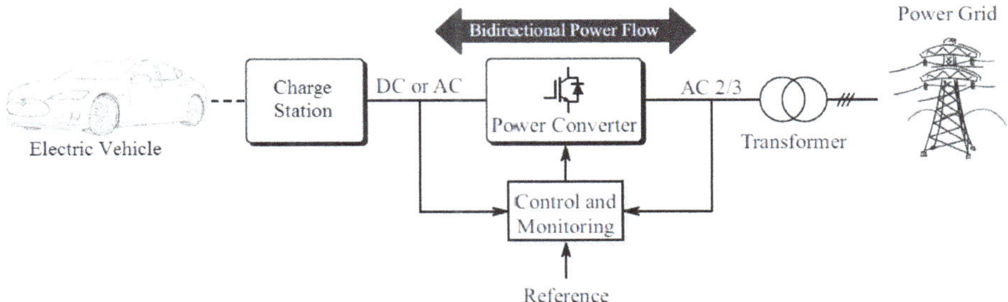

Figure 4. Block diagram of the V2G structure [22].

2.2. Vehicle to Home (V2H)

This technology refers to the systems in which the energy stored in batteries of the electric vehicles can be used as an energy source for houses [35]. For example, during the night, when power consumption is low on the grid, the battery can be charged and the stored energy can be sold to the grid when the energy consumption is high on the grid [36]. The V2H block diagram is shown in Figure 5. The electric vehicle is plugged into a charging station from where it gets energized. An energy management system and a home load manager are connected in parallel to the vehicle. The controllable switch before the transformer with the main grid controls whether power will be exchanged with the main grid or not. When the switch is open, the vehicle transfers energy to the home loads. The energy management system serves to supervise the energy transfer to ensure that everything is working normally. The size for a battery energy storage system in residences can vary from 3 to 30 kWh depending upon the manufacturer [37]. Therefore, V2H technology can play a significant role in exploiting EV energies to power up residential loads.

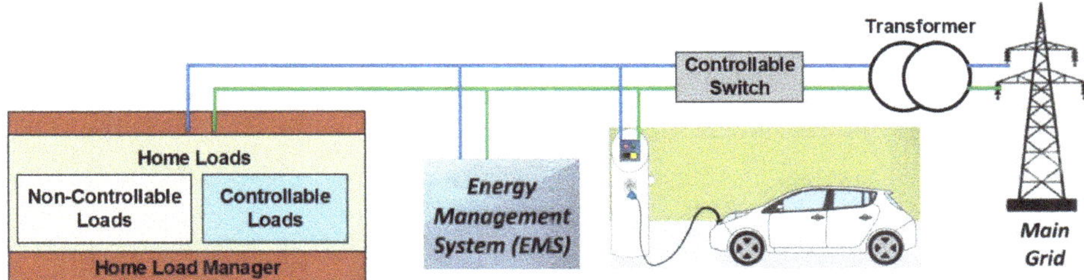

Figure 5. Block diagram of the V2H structure [36].

The infrastructure of V2H technology is similar to the V2G structure. The main activity in this case is to transfer energy from vehicles to houses, buildings, or other electrical vehicles in line with the existing energy capacity [38]. However, there are critical conditions for energy transferring. The most important of these conditions is the amount of energy.

As depicted in Table 1, if the energy is to be transferred to a detached house from an electric vehicle, the amount of power in the electric vehicle must be between 5 and 10 kW. In case of transferring from an electrical vehicle to a building, it must be between 10 and 15 kW. Or, if the energy is transferring from an electrical vehicle to another electrical vehicles, this value must be between 15 and 30 kW [39]. These standards are very important for energy efficiency, continuity, and reliability [40].

Table 1. Vehicle-to-X energy transfer range [39].

Power Flow	kW	Vehicle-to-X	Objective
Bi-directional	5–10	Vehicle-to-home (V2H)	It is used in case of emergency energy demand and for storage energy in the battery
Bi-directional	10–15	Vehicle-to-building (V2B)	It is used in case of emergency energy demand
Bi-directional	15–30	Vehicle-to-grid (V2G)	It is used to participate in the large scale energy market

3. Charging Systems of the Batteries of Electrical Vehicles

It is very important to store as much of the energy produced as possible. V2G technology works with specially designed bidirectional charging stations that allow the electrical vehicle owners to charge their vehicles while discharging the vehicle battery. The electricity in the batteries of electrical vehicles is transferred to the grid to compensate for the supply–demand balance in the electricity grid [40].

The ability of the batteries to charge and discharge depends on many factors, such as the design of the batteries, their charge status, temperature, former cycle history, and use. Depending on the charging strategies and charger size of the electric vehicle batteries, the peak power demand of the grid can vary [41]. This multiple dependency makes the determination of the charge status of the battery and the charging methods complicated. The battery charging methods used in the literature are constant current charging, constant voltage charging, and constant current–constant voltage charging. The charging current in the constant current scheme is equal for all battery groups that are connected in a series. As the charge status increases in batteries, the voltage must be increased in a constant manner to continue charging at constant current as the internal resistance also increases [42]. However, the charging current to be selected is very important in this method. This is because a high value of charging current allows the battery to be charged in a short time; however, it also causes damage to the battery because of overcharging and overheating. Charging the batteries at low current increases the charging times. At constant voltage, the battery charge draws a high current at the initial stage from the source because of the low battery internal resistance. This high current is limited to avoid damage to the elements. In constant voltage charging, the charging is started at full current of the charger by applying voltages that cannot cause damage to battery elements. After reaching the voltage level, called the float voltage, the current gradually begins to decrease. The charging current decreases in time because of the increasing battery internal resistance that stems from the increase in the charge [43]. This allows the charge to be completed with the leakage current, and in this way, the possibility of overcharging the battery is reduced. Because of the reduction in the charging current, the charging time of the battery becomes longer. Constant current–constant voltage charging is applied in two steps. For the purpose of eliminating the negative conditions, such as the overcharging of the batteries and pulling the overvoltage from the batteries, the battery is charged with a constant current until it reaches the preset voltage level; then, charging is continued with the constant voltage level [44].

The discharge levels, temperatures, and charge method parameters of the batteries of electrical vehicles affect the battery life cycle. To protect the battery life cycle, it is necessary to have a charging topology with a high efficiency or to select proper charging topology befitting the characteristics of the battery. The most important characteristics of these charge topologies are to provide the proper voltage level according to the energy flow direction and bidirectional energy flow [45]. Also, the methods are standard for battery charging of electrical vehicles. There are three main charging methods, named Type-1, Type-2 and Type-3. These are classified on the basis of their usage and applications. Type 1 charging method is used for vehicles, which are usually parked in residences and workplaces for a long time because of its single-phase system. The battery charging time being long and slow does not cause overload to the existing grid. For this reason, overnight charging is carried out to benefit from cheap electricity. When the battery is full, it provides power up to 3.7 kW and a maximum current of 16 A in Type 1 charging mode [46]. Type 2 charging method is used in places where there is heavy density, such as hotels, markets, hospitals, universities, airports, and shopping centers. It provides medium-speed charging within 1–4 hours of periods. It has a three-phase AC grid, and provides power between 11 and 22 kW, and a maximum current of 32 A [46]. Type 3 charging method is also defined as the method of fast charging. The battery charging time varies between 15 and 30 minutes, and these stations offer the possibility of charging batteries within short times in areas such as short breaks where there is an urgent need for energy. Although it has both the AC and DC model, it causes too much load for the network because of its high current value. The technical specifications of the charging stations used to charge batteries of EVs are categorized according to the type of the charging stations. They provide powers up to 43 kW for AC, and the maximum values for DC are 500 V and 125 A. In Type 3 charging method, there are safety measures present, such as the verification of the cable connection, not giving voltage when the cable is not connected, checking the ground connection, and reporting the maximum current capacity of the charger [47,48].

Overcharging of the batteries causes disruptions in the chemical structure and shortens their usage of life cycle. Charging systems have to work together with the battery management system to avoid overcharging. In addition, energy management systems are needed to ensure that batteries can be used safely under normal operating conditions and even in the event of accidents. The basic functions of battery management systems are to provide protection for the cells, heat management, charge/discharge control, data collection, communication with modules, data storage, and cell balancing [49]. In the case of the battery of the vehicle wearing out, the battery becomes eligible for a second-time use. The batteries are known as 'second life batteries', of which their major source is EVs. Such batteries can be repurposed for use in residences, telecommunication towers, building loads, and in power and transmission support [50]. They are very much cost-effective for their use in residential areas for following loads and backing-up the systems, the same as in other commercial and industrial areas [51].

The general energy flow diagram of charge/discharge processes of V2G and V2H structures is given in Figure 6. Different AC/DC and DC/DC topologies are used to increase the efficiency and performance of this system. There is a need for the control system to manage the energy flow, and, for this reason, all these topologies are employed in this respect. The charging time of electrical vehicles is longer when compared with fuel filling time. Fast charging methods are employed to shorten this time. In this method, the energy flow-control is applied up to 80% of the battery, and voltage control is applied over 80% [52].

The Li-ion battery pack is the battery of the vehicle, which is connected to a charger module. The grid supplies power to other loads and, by means of a suitable transformer to alter the magnitude, to the charger module. The charger module has two converters, firstly an AC/DC converter to convert the AC grid power into DC required by the battery, and secondly a DC/DC converter to change the magnitude of the DC power as necessary. Both these converters are controlled by the mechanism of pulse width modulation (PWM). This system is bidirectional and can work in either V2G or G2V technologies [52]. Li-ion batteries used in electrical vehicles have approximately 5000 life cycles. One charging and one discharging of the battery constitutes a cycle [53]. In Figure 7, when the number of cycles increases, the energy holding capacity of the battery decreases. Reduction of capacity also means shortening of battery life. A battery management system is needed to increase the operating time [54].

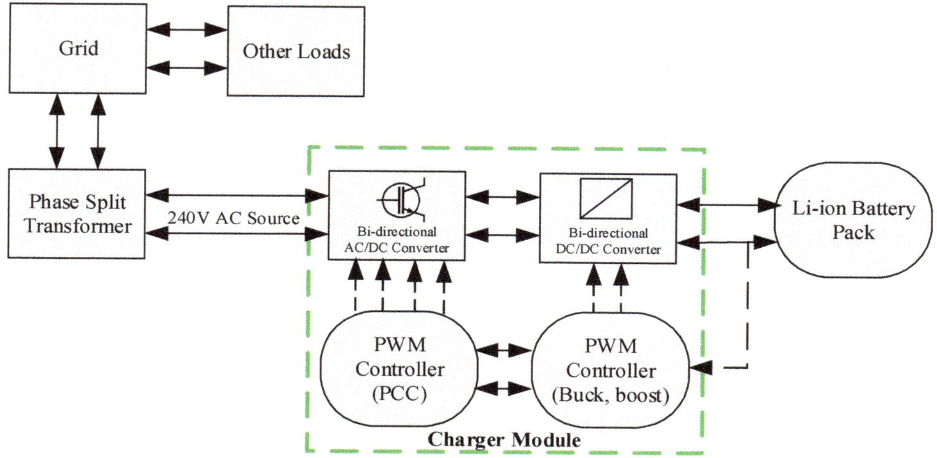

Figure 6. Generalized energy flow diagram for V2G and V2H system [52]. PWM: pulse width modulation.

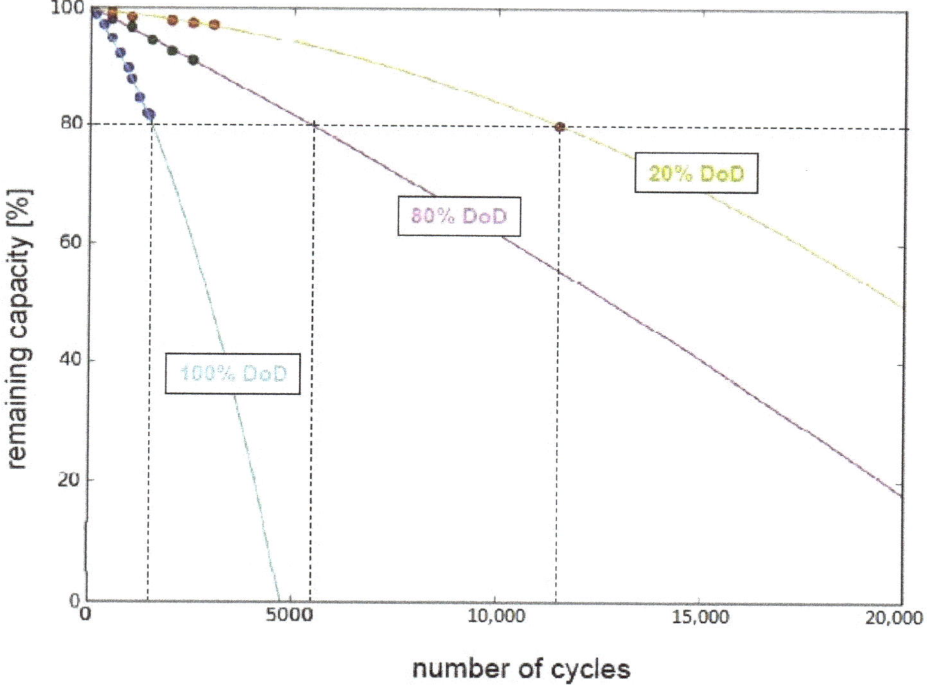

Figure 7. Li-ion battery usage curve: the energy storing capacity of the battery is found to drop with increasing number of charging/discharging cycles [55].

Determining the charging locations has fomented significant research interest among scientists and engineers. Optimum charging points are determined using various heuristic algorithms such as the genetic algorithm [56], non-linear auto regressive [57], flow refueling location model [58], maximum covering location problems [59], and agent-refueling multiple-size location problem [60]. This is usually important for V2B applications, as the buildings are densely located and the intermittency of loads are high. Queuing algorithms are efficient in such cases, wherein the charging stations need to balance the loads to minimize the charging time [61,62]. In such stations, the chargers are specified in three levels on the basis of their charging power and charging circuit [63]. Queuing models according to these levels helps to design the stochastic resource-sharing network to accommodate the convoluted distribution [64] and traffic [65,66] networks to make V2X more accessible.

Battery management systems (BMS) are important components in the provision of conditions such as safe operation, long-term reliability, and low cost of a battery. BMS increase battery life cycle and prevent damage to the battery, ensuring correct and reliable operation of the system [67,68].

The curve showing the relationship between temperature and battery output voltage is depicted in Figure 8. Accordingly, the battery temperature increases when the battery draws excess current. Therefore, the increased temperature also adversely affects the output voltage [69]. For this reason, the current is required to be limited to a certain value when the battery is in constant current mode. The limitation process is performed through BMS.

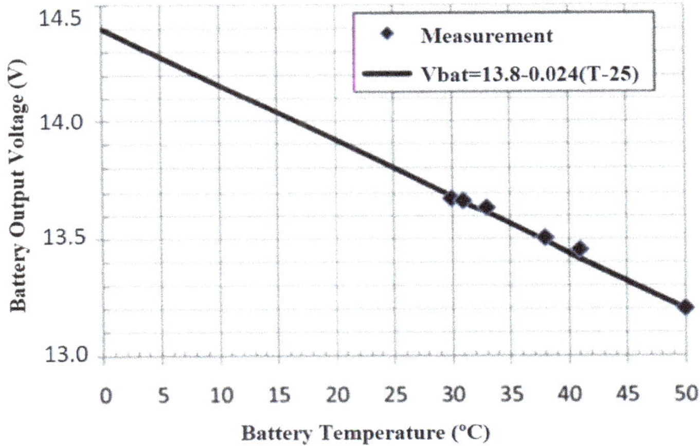

Figure 8. Relationship between temperature and battery output voltage: the output voltage reduces linearly with increasing battery temperature [69].

The BMS provide the balance of the voltage values of each cell that makes up the battery pack in order to maximize the capacity of the batteries and to prevent overcharging when charging [70]. In case of over voltage or under voltage in any cell, the system interferes in case of unbalanced voltage balances and the system enters the cutting [71]. When necessary, it provides balance by transferring the energy from the most filled cell to the least charged cell. In this way, BMS intervene in the system and prevent damages in the statement of system failure by interrupting. This is an extremely important system for the protection of high capacity and high cost battery packs [72].

BMS provide the protection of the system by interfering with the system when optimal values are exceeded, done by measuring the values presented to the user. BMS interfere with the high current which is drawn from batteries and interfere with the system during the high charge, the low voltages during discharge, the high temperature, the low temperature, and the leakage current formation. [73].

3.1. Various Converter Topologies Used in V2G and V2H Technologies

3.1.1. Isolated and Non-Isolated Converters

The magnitude and type of electrical energy changes according to the flow direction of the energy in the battery or in the grid. Power electronics topologies are employed for this change.

The bidirectional AC/DC power converter topologies used in the charging–discharging systems of batteries in V2G and V2H technology are classified as illustrated in Figure 9. The converters used for charging or energizing the grid operate in a bidirectional way. Converters have advantages as well as disadvantages in terms of quality of energy compared to preferred topologies [74].

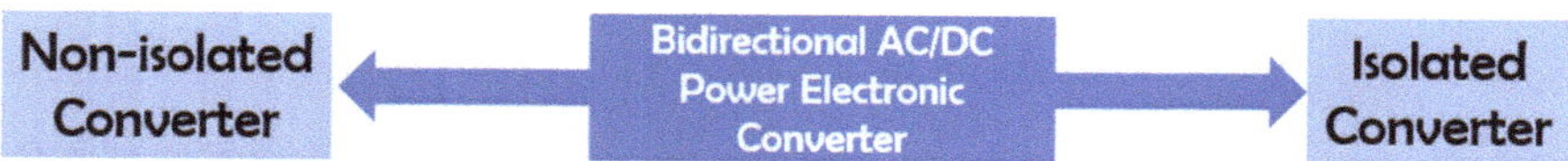

Figure 9. Classification of bidirectional AC/DC power converter topologies: the bidirectional AC/DC power electronic converters are mainly of two types, isolated and non-isolated converters.

Non-isolated converters are obtained in structural terms by connecting an active and a passive semiconductor switching power element and an inductance in different ways. The operating mode of inductance converters is based on the transfer of energy that is stored in the inductor. As long as the semiconductor switching power element is actively transmitting, the energy that is provided by the source that is stored in the inductor being transferred to the cut-off by the semiconductor switching element, which results in the transfer to the load. The important disadvantage of these converters is the lack of isolation between the output and the input [75].

Isolated converters are used in situations when the electrical isolation is required in DC/DC transducer applications or where there is a high rate between input and output. Here, a transformer is used to provide isolation. In essence, the working principle of the isolated transformers is the same as the non-isolated converters. In other words, it is based on the logic of transferring the energy that is stored in the inductance. As long as the semiconductor switching element is actively transmitting, the energy that is provided by the source and stored in the inductor is transferred to the cut-off by the semiconductor switching element, which results in the transfer to the load.

The DC/DC or DC/AC power converters that are employed in charging topologies are usually controlled with two different methods, these being pulse width modulation (PWM) and frequency modulation (FM). In the FM technique, the output value is controlled by changing the pulse frequency of the semiconductor switching element, in other words, by changing its period [76]. However, this technique is mostly used in compulsory situations such as in temporary and low load situations. In addition, as a result of using this technique, fluctuations and noises occur in the output voltage. The PWM technique is widely used in industrial applications as it allows filtering of the fluctuations and noises at the input and output and because of its constant frequency operation. The PWM technique is a method in which the output value is controlled by adjusting the operating time of the semiconductor switch by changing the pulse width at constant frequency. Here, the determination of the operating time of the key by producing a control signal that is needed for the semiconductor switching element with PWM is shown in Figure 10 [77].

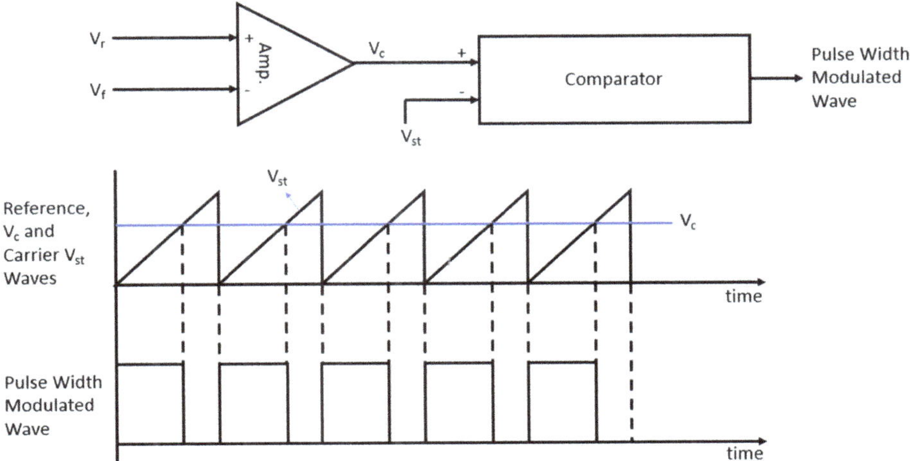

Figure 10. Pulse width modulation technique: a pulse width modulated wave is found by comparing a reference signal, Vc, and a carrier signal, Vst.

3.1.2. Bi-Directional Half-Bridge and Full-Bridge Controlled Converter

In Figure 11, the circuit diagram of a bi-directional half-bridge and full-bridge controlled converter is given. In the bidirectional half-bridge-controlled converter, the AC energy received from the electricity grid is converted into DC energy in half-bridge with the switching elements and is reduced to the voltage level of the battery by using the converter (buck converter). At the same time, the DC received from the battery is amplified by the boost converter circuit, which increases the voltage and then gives it to the grid by converting it from DC to AC through diodes [78]. In the bidirectional full-bridge-controlled converter, the conversion process from AC to DC is done through a full-bridge converter circuit. The other parts are the same as the bidirectional half-bridge-controlled converter [78,79].

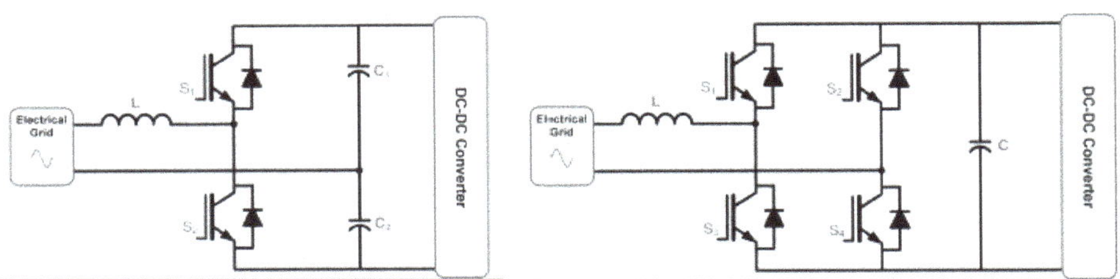

Figure 11. Bi-directional half-bridge and full-bridge controlled converter.

3.1.3. Bi-Directional Buck–Boost Isolated Converter

The charge level of the battery is required to be equal or more than the voltage converted from AC to DC. Similarly, it is necessary that the current level drawn from the battery is adequate in the conversion process. If these conditions do not exist, voltage collapse and similar negative situations occur. For this reason, the bi-directional buck–boost converter topology is needed in power electronics. In Figure 12, a bi-directional buck–boost converter and bi-directional isolated converter circuits are depicted. In the conversion process, the S1 switch is used in buck operation, and the S2 switch is used in boost operation [80]. Thus, the desired voltage level for the battery charge and the voltage levels needed in the battery to convert the energy to AC are obtained. In the bidirectional isolated converter, the DC voltage converted from the AC is transferred to the other side through an AC voltage and isolated transformer; then, it is converted into DC to charge the battery. The same processes also apply in transferring from the battery to the grid. Firstly, the voltage that is obtained from the battery is converted into AC and then transferred from the isolated transformer to the other side, thereby being converted into an alternating current through the AC voltage corrector circuit. Here, the grid and the battery part of the circuit are isolated with the isolated transformer circuit to ensure circuit protection [80,81].

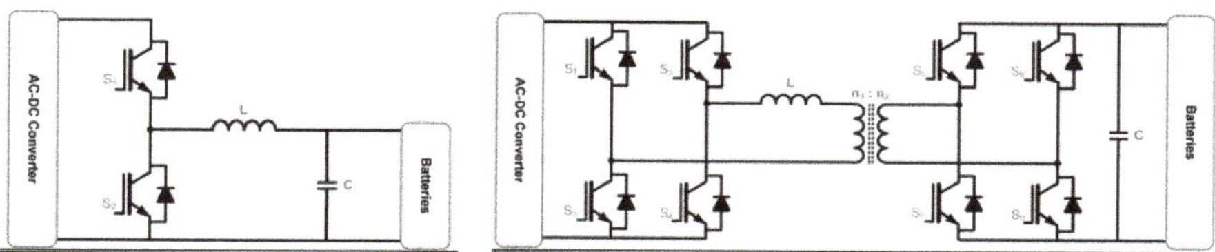

Figure 12. Bi-directional buck–boost converter and bi-directional isolated converter circuits.

3.1.4. Non-Isolated Charging Topology with PWM and Bi-Directional Buck–Boost DC/DC Converter

In Figure 13, the non-isolated charging topology, which consists of PWM (pulse width modulation) and bidirectional buck–boost DC/DC converter are shown [82]. Firstly, the AC grid signal is converted into DC voltage by the converter circuit and is filtered by the capacitor; then the battery is charged by using the S5 switch (reducing converter). Similarly, the DC voltage obtained from the battery is increased by the switch S6 (increasing converter), and converted into AC by the inverter circuit and given to the grid [83–85].

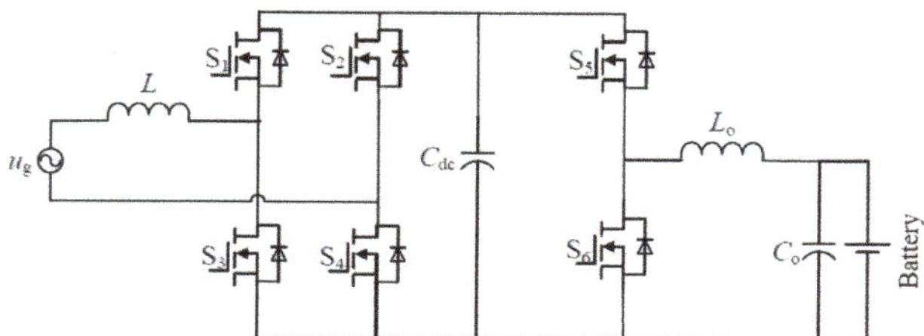

Figure 13. Non-isolated charging topology with PWM (pulse width modulation) and bidirectional buck–boost DC/DC converter.

3.1.5. The Non-Isolated Charging Topology with PWM and Bi-Directional Cascade DC/DC Buck–Boost Converter

Figure 14 illustrates the non-isolated topology, which consists of PWM (pulse width modulation) and the bidirectional cascade DC/DC buck–boost converter. Firstly, the grid signal is converted into DC voltage by the rectifying circuit and is then filtered by the capacitor and the coil. Then, the battery is charged by using the buck–boost converter circuit. Similarly, the DC energy obtained from the battery is increased with a converter circuit, which increases or decreases the voltage to the grid, and is then converted into AC voltage by the inverter circuit [86].

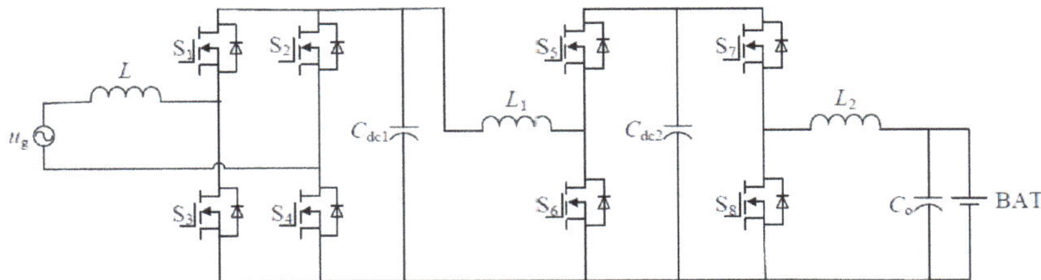

Figure 14. The non-isolated charging topology with PWM and bidirectional cascade DC/DC buck–boost converter.

3.1.6. The Two-Stage Topology with PWM Convertor—Active Double Bridge and Series Resonance Convertor

In Figure 15, the bidirectional topology is highlighted, consisting of a PWM converter and active double-bridge. In Figure 16, a two-stage topology is given, consisting of a PWM converter and series resonance converter. In both topologies, the grid and battery sides are isolated by using an isolated transformer [87,88]. In Figure 16, a capacitor is used in the topology, which consists of a series resonance converter, to increase the output voltage and efficiency [20,89]. Full-bridge AC/DC converter with PWM controllers are also widely used in different switching power converters [90,91].

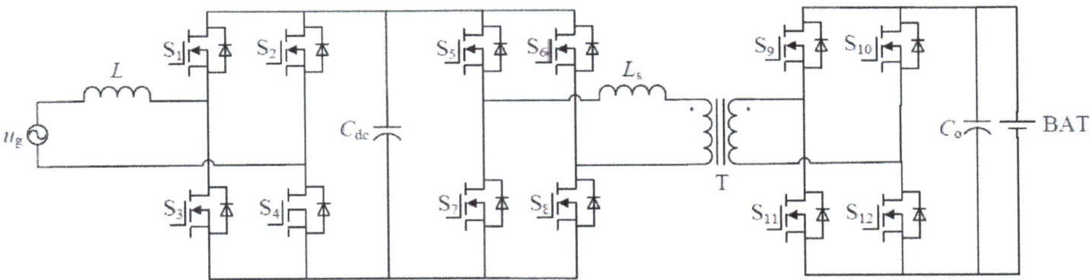

Figure 15. The two-stage topology with PWM convertor and active double bridge converter.

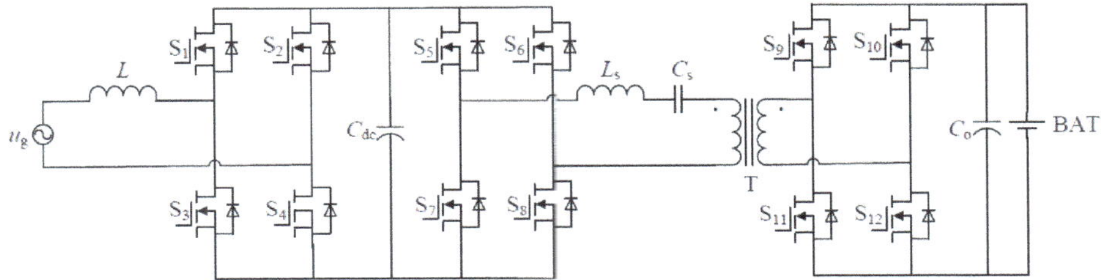

Figure 16. The two-stage topology with PWM converter and series resonance convertor.

3.1.7. Buck–Boost DC/DC Convertors

Buck–boost converters essentially consist of non-isolated type power converters, which consist of a functional combination of a buck converter and a boost converter. The factor, which determines whether such convertors work as buck or boost convertors, is determined by the duty rate (D), which is the rate of the pulse width to the total period [92]. When the semiconductor switching element in its structure is in the transfer position, it is fed only by the inductance, and in this way, the current passing through the inductance increases in a linear way, and the energy level of the inductance is increased [93]. The feeding of the load is provided by a capacitor. When the semiconductor switching element is in the cut-off position, the power diode starts transmission, and the output is fed by the energy that is accumulated in the inductance [94]. After this point, the inductance current decreases in a linear way, and the energy level of the inductance is reduced. Here, the power elements are exposed to the total of the input and output voltages [95]. In addition, since the direction of the output voltage is reverse to the input voltage direction, these converters are also known as inverted convertors [10]. The circuit structures, which show the basic circuits of the semiconductor switching element and the transmission and cutting status of the semiconductor switching element and basic waveforms, are given in Figure 17.

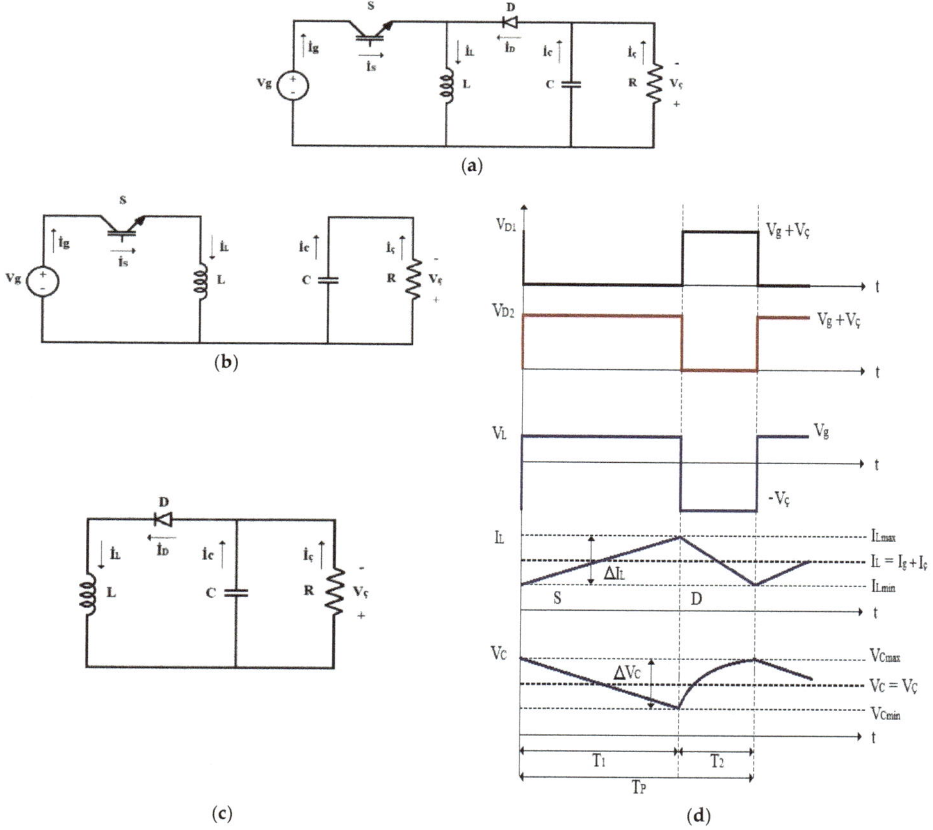

Figure 17. (**a**) Basic circuit, (**b**) status with turned-on IGBT, (**c**) status with turned-off IGBT, and (**d**) waveforms of the circuit of the buck–boost convertor.

4. Communication Standards in Electrical Vehicles

Interactive operation with the grid of electric vehicles and the communication methods used in the charging mode of electric vehicles cause a complex structure [96]. Seamless communication is required to design charging stations with sharing networks. To effectively schedule charging operation for the users, improved and resilient communications are imperative [97]. Various internet-based communication schemes [33,98–100] have been proposed to avoid the compatibility issues among charging stations. Furthermore, to prevent this negative situation, communication standards have been established, and it has become compulsory for the production companies to comply with these standards. There are four sets of standards considered for EVs: (1) plug, (2) communication, (3) charging topology, and (4) safety. All these standards are maintained in the V2G technology. These standards that apply to the charging of electrical vehicles are illustrated in Figure 18. The connector structures, the communication methods, the charging topologies, and the safety and the interoperability standards are given separately [101].

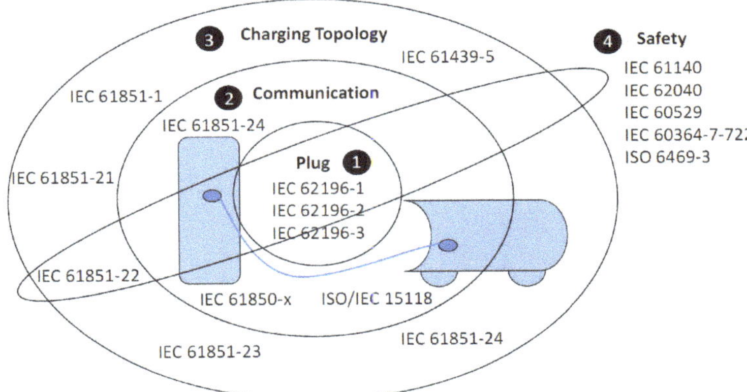

Figure 18. Standards on charging processes of electric vehicles.

In Figure 19, the sample communication standards of the vehicles using V2G technology are given [102–108].

- IEC 62196-1: Plugs, socket-outlets, vehicle couplers and vehicle inlets—conductive charging of electric vehicles, charging of electric vehicles up to 250 A AC and 400 A DC.
- IEC 62196-2: Plugs, socket-outlets, vehicle connectors and vehicle inlets—conductive charging of electrical vehicles, dimensional compatibility, and interchangeability requirements for AC pin and contact-tube accessories.
- IEC 62196-3: Plugs, socket-outlets, and vehicle couplers—conductive charging of electric vehicles, dimensional interchangeability requirements for pin, and contact-tube coupler with rated operating voltage up to 1000 V DC and rated current up to 400 A for dedicated DC charging.
- IEC 61850-x: Communication networks and systems in substations.
- ISO/IEC 15118: Vehicle-to-grid communication interface.
- IEC 61439-5: Low-voltage switchgear and control gear assemblies, and assemblies for power distribution in public networks.
- IEC 61851-1: Electrical vehicle conductive charging system—general requirements.
- IEC 61851-21: Electrical vehicle conductive charging system—electric vehicle requirements for conductive connection to an AC/DC supply.
- IEC 61851-22: Electrical vehicle conductive charging system—AC electric vehicle charging station.
- IEC 61851-23: Electrical vehicle conductive charging system—DC electric vehicle charging station.
- IEC 61851-24: Electrical vehicle conductive charging system—control communication protocol between off-board DC charger and electrical vehicles.
- IEC 61140: Protection against electric shock—common aspects for installation and equipment
- IEC 62040: Uninterruptible power systems (UPS).
- IEC 60529: Degrees of protection provided by enclosures (IP code).
- IEC 60364-7-722: Low voltage electrical installations, requirements for special installations, or locations—supply of electric vehicle.
- ISO 6469-3: Electrically propelled road vehicles, safety specification, and protection of persons against electric shock.

Both data and energy flow are bidirectional among the vehicles, charging stations, and the grid. The ISO/IEC 15110 standard is used for communication between the EV and the charging station, whereas the IEC 61850 standard is used for communication between the charging station and the grid. In addition, the communication among the charging stations and the smart grids are carried out to facilitate charging and supply tariffs dynamically [109–111]. EV fleeting operators (FO) are being highly recommended by researchers to utilize the new business opportunities by giving varying services to system operators [112,113].

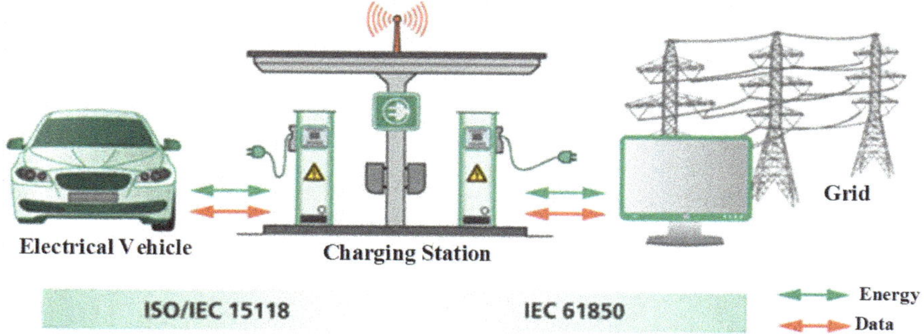

Figure 19. Sample communication standard of the vehicles using V2G technology.

As seen in Figure 20, when the connection socket is attached between the grid and the electrical vehicle, the identification, the authorization, and the verification procedures are done to ensure a secure connection.

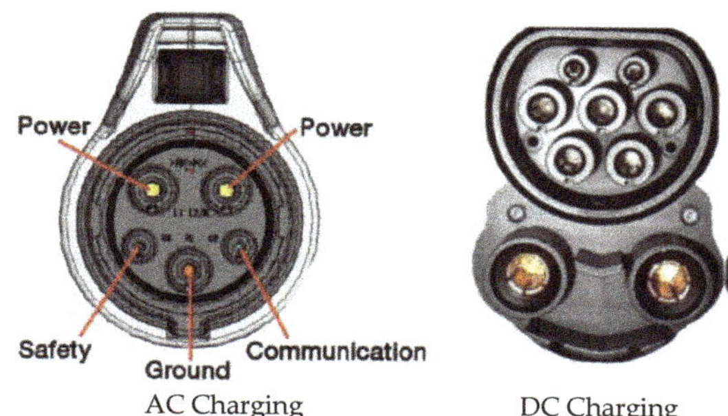

AC Charging   DC Charging

Figure 20. AC and DC charging socket structure. These sockets are attached with the vehicle and the grid for the transfer of energy between them.

5. Discussion

The bidirectional AC/DC converter topologies that are employed for V2G and V2H technologies and their features, advantages, and disadvantages are given in Table 2. In general, full-bridge topology is preferred in AC/DC conversion. As the power of the system increases, the number of the semiconductor switches that are employed in the switching process increases also, which, in return, increases the total harmonic distortion (THD) of the system while decreasing its efficiency. The output voltage of the converter varies depending on the battery voltage level or DC bus voltage (G2V or H2V). LC or LCL filters are preferred for connection to the grid in the V2G direction. These filters reduce ripples in the current and voltage.

Table 2. Comparison of bi-directional AC/DC inverter topologies for V2G applications in the literature.

Bi-Directional AC/DC Inverter Topology	Power	DC Bus Battery Voltage	Number of Switching Devices	Filter	Power Factor	Total Harmonic Distortion (THD)	Advantages/ Disadvantages	Ref. Number
Full bridge	500 W	60–120 V	4 IGBTs 4 MOSFETS	LC	1	not known	Low efficiency Hard switched	[114]
Full bridge	500 W	60–120 V	4 IGBTs	LC	0.99	4.3%	No DC bus capacitor	[114]
Full bridge	3.5 kW	300–340 V	4 IGBTs	RLC	1	not known	No isolation	[115]
Full bridge	3.6 kW	270–360 V	4 IGBTs	L	0.99	<3%	Low THD	[116]
Full bridge	3 kW	120 V	4 MOSFETS	-	1	4.5%	High THD	[1]
Full bridge	3.3 kW	400–450 V	4 IGBTs	LCL	variable	not known	Fast response, compensation	[117]
Full bridge	400 W	120 V	4 MOSFETS	LC	variable	6.15%	High THD, compensation	[118]
Three phase full bridge	20 kW	800 V	6 IGBTs	LCL	not known	3 %	99% efficiency (due to SIC devices)	[20]
Three level	18 kW	350 V	8 MOSFETS	-	1	2.3%	Low THD, more number of switches,	[119]
Single state isolation	3.3 kW	280–350 V	8 MOSFETS	3 LC	0.98	<5%	97% efficiency, complex control	[120]

The bidirectional DC/DC converter topologies used for V2G and V2H technologies and their properties, advantages, and disadvantages are given in Table 3. When the studies given in Table 3 are considered, for the increase of the voltage level for V2G or V2H transfer, or for the reduction of the voltage level for G2V transfer, the bidirectional buck–boost inverter is preferred. In the comparison table, the system feedback of the closed-cycle control system is employed with a current-controlled setting, voltage-controlled setting, or both together. As the battery voltage level increases, high-frequency switching is needed to control the system, and the efficiency of the system reduces. As the magnetic isolation between the inverter and the converter reduces the current and voltage ripples that are caused by the high-switching frequency, it increases the whole system efficiency.

Table 3. Comparison of bi-directional DC/DC converter topologies for V2G applications in the literature.

Bi-Directional DC/DC Convertor Topology	Type of Feed	Power	DC Bus Battery Voltage	Number of Switching Devices	Filter	Isolation	Switching Frequency	Efficiency	Advantages/ Disadvantages	Ref. Number
Buck–boost converter	Current–voltage feed	500 W	60–120 V	2 IGBTs	C	No	20 kHz	83%	Low efficiency, high current ripple	[121]
Buck–boost converter	Dual current feed	500 W	60–120 V	2 IGBTs	LC	No	10 kHz	<85%	Low efficiency, high current ripple	[122]
Buck–boost converter	Dual current feed	3.5 kW	270–360 V	2 IGBTs	LC	No	20 kHz	>90%	Fewer components	[123]
Buck–boost converter	Current–voltage feed	1.2 kW	100–130 V	2 IGBTs	LC	No	50 kHz	Not known	Fewer components	[124]
Interleaved buck–boost converter	Current–voltage feed	30 kW	170–200 V	4 IGBTs	C	No	20 kHz	Not known	High power transfer	[125]
Interleaved buck–boost converter	Current–voltage feed	400 W	120 V	4 IGBTs	C	No	20 kHz	>94%	Low power output	[117]
Cascaded buck–boost converter	Dual current feed	9 kW	350 V	4 IGBTs	LC	No	20 kHz	91%	High transient stability	[126]
Dual full bridge converter	Dual voltage feed	3.3 kW	230–430 V	8 IGBTs	CLC	Yes	250 kHz	Not known	High frequency	[96]
Dual full bridge converter	Dual voltage feed	30 kW	360 V	8 IGBTs	C	Yes	20 kHz	Not known	Compensation	[120]
Half full bridge converter	Current–voltage feed	1 kW	250–450 V	6 IGBTs	C	Yes	100 kHz	95%	High efficiency, control flexibility	[127]

Strengths of V2G and V2H technologies

- V2G and V2H technology ensures that the reactive power is compensated by providing active power or renewable energy sources in the existing grid.
- The advantages of V2G and V2H technology are not only possible for the grid, but also for the owners of electrical vehicles. These systems provide vehicle owners with continuous power support at home or at work.
- Increasing the capacities of the existing energy sources or preparing new energy sources necessitates high costs; therefore, it is less costly to have support from V2G or V2H technologies in periods when the demands are high.
- These technologies increase the energy quality, reliability, and sustainability by reducing frequency regulation and harmonic distortion.
- The technologies are compatible with micro grid and smart grid applications.
- Electrical vehicles provide more stable, safer, and more continuous energy backup or emergency energy support compared with solar wind and other renewable energy sources, which depend on charging.
- While electrical vehicle owners charge their vehicles at low-cost rates at night, they sell energy to the grid during peak hours when energy is expensive within the day, and, thus, obtain financial gains.

Weaknesses of V2G and V2H technologies

- The life cycle of the batteries will shorten as the charge–discharge process will increase the internal resistance considerably. This negative situation is considered as the disadvantage of these technologies.
- As the fast charging method also shortens the life cycle of batteries, the use of such technologies is not recommended because they cause the breakdown of batteries.
- Purchase of electric vehicles that have V2G or V2H technologies requires high initial costs.
- Coordination and standardization with the grid operators are difficult at initial steps.

Opportunities of V2G and V2H technologies

- The battery management system can be formed by using optimization and control algorithms to extend the service life of the battery.
- Software and hardware may be developed to measure battery health status to estimate the service life of batteries. In this way, the owner of the battery can be informed before the battery life ends, and thus measures can be taken in terms of contributing to the continuity of the energy.

Threats of V2G and V2H technologies

- As cyber-attacks are becoming increasingly complex, providing necessary measures to deal with current cyber threats will not provide adequate protection. The power system may also become vulnerable to new attacks in the future. For this reason, it is necessary that the basic components of the power system are defined and protected as a whole.
- Wired or wireless communication methods are employed to ensure the security between the energy systems, grids, electric vehicles, and charging station, which are among the critical infrastructures. For this reason, a possible cyber-attack to these critical communication methods may damage the whole system. Necessary preventions must be taken in this respect.
- All batteries lose their capacity over time. Therefore, the amount of energy to be transferred to the grid and the energy to be sold to the grid will decrease with time.

6. Outcomes

The purpose of this paper is to summarize and arrange all necessary information about V2G and V2H technologies, along with their communication standards and charging topologies, such that it provides a substantial knowledge to a beginner in this field. The key findings of this paper are listed below:

- The world's concern for the environment is on the rise, as traditionally-used non-renewable fuels are harmful and expensive. In order to ensure green transportation technologies, the concept of electric vehicles (EV) has come into the limelight. EVs run on electricity, posing no threats to the environment. They can also be developed on the basis of existing electricity infrastructures, making them less costly. Their additional advantage is that they can store electrical energy and can act as a source when not in use. This feature of EVs has ushered the dawn of V2X technologies.

- V2X is a general term where X is a variable representing either grid (G), home (H), device (D), building (B), or vehicle (V). These are technologies wherein electric energy is supplied from vehicles to the grid, a home, a device, a building, or to another vehicle.

- Vehicle to grid (V2G) technology is a bidirectional energy transfer between a vehicle and the electricity grid. The energy transfer also includes necessary AC/DC or DC/AC conversion, along with magnitude changing. The vehicle charges itself from the grid when the electricity demand is low, or when the electricity prices are less. On the other hand, the vehicle discharges or supplies energy to the grid during the peak demand hours when the electricity prices are high. Thus, the vehicle owner obtains a financial profit through this technology.

- Vehicle to home (V2H) technology is similar to the V2G technology, except that the energy transfer is between a home and the vehicle. If a vehicle supplies a house with energy during the peak hours, the demand on the grid reduces, making the electricity distribution smoother. Again, the vehicle can get charged from the off-peak hours.

- The charging systems of the batteries of the EVs depend on multiple factors, such as the design of the batteries, their charge status, temperature, former cycle history, and usage. There are two main charging systems, namely constant current charging, and constant current–constant voltage charging. The constant current charging often results in overcharging the battery, thereby overheating it and damages it. On the other hand, the constant current–constant voltage method eliminates the risk of overcharging but increases the charging time of the battery. As an optimization, the battery is charged at a constant current until a preset voltage is reached, and then charged at a constant voltage. A battery management system (BMS) is employed in the system to act as an overall controller of the battery health by monitoring its charge status, temperature, battery cycles, and other such parameters, providing data to and communication with other modules.

- The bidirectional AC/DC power converter topologies used in the charging–discharging systems of batteries in V2G and V2H technology are classified into isolated and non-isolated converters on the basis of the presence or absence of a transformer between input and output to provide isolation. The converters use either pulse width modulation (PWM) or frequency modulation (FM) to control the output voltage.

- The bidirectional AC/DC converters can be either half-bridge or full-bridge. In both cases, buck and boost converters are employed to reduce or enhance the voltage level.

- The voltage level of the receiver must be less than that of the supplier. Hence a definite voltage level must be maintained in the vehicle's battery in order to supply energy to or extract energy from the grid. Otherwise, a definite current level must be maintained to protect the battery, along with ensuring fast charging. These conditions are met using a bidirectional buck–boost isolated converter, which can alter the voltage levels as and when required.

- A non-isolated charging topology with PWM control can also be employed with bidirectional DC/DC buck–boost converters to raise or lower the voltage levels.

- Multiple units of DC/DC buck–boost converters can also be cascaded to perform multi-step reduction or amplification of voltage in case of the non-isolated charging topology with PWM control.
- The charging topology can also be composed of two-stages isolated by a transformer. The two stages can be formed either with a PWM converter and an active double bride converter, or with a PWM converter and a series resonance converter.
- Buck–boost converters are extensively employed in the V2G operation to obtain the desired voltage level in various stages of the energy transfer process.
- For a successful and organized transfer of energy in the V2G technology, a good communication is required between the vehicle and the grid operator. There are predefined communication standards set for this purpose. The standards vary according to the connector structures, the communication methods, the charging topologies, and the safety and the interoperability standards.
- In the V2G technology, both AC/DC and DC/DC converters are indispensable. There can be numerous topologies of these converters. For each type of AC/DC converter, the power consumption, power factor, number of switches, type of filter, merits and demerits, and THD are explored. Similarly, for the various types of DC/DC converters, the type of input, power consumption, efficiency, switching frequency, and merits and demerits are examined.
- Finally, an extensive SWOT analysis of the V2G technology is made, wherein it is evident that, despite the few weaknesses and threats, this emerging technology has a promising future and can contribute towards building a greener and much more efficient energy infrastructure.

The charging topologies and the communication standards of V2G and V2H technologies can be enriched further with more study, research, and development in this sector. There can be better charging topologies that will be efficient as well as inexpensive. The communication standards can be made more reliable and versatile so that the need for so many individual standards for individual purposes is quenched. Future research work in this field can be directed towards this motive.

7. Conclusions

This study is an assisting document for those who want to work in this field. With the help of the technology developed in recent years, electrical vehicles now have access to V2X technology. Electrical vehicles are useful for preventing greenhouse gas emissions and global energy and climate changes, and promoting the efficient use of energy and energy saving. V2G, V2H, or V2X technologies provide an efficient use of existing energy resources and also reduce the infrastructure costs of the planned energy sector. The continuity of the existing energy in the grid is ensured with V2G or V2H technologies, together with an efficient and reliable source formation, a stable operation, and a high quality of power. In the future, the problem of emission from vehicles will be eliminated completely by bringing the grid power connection standard in the electrical vehicles. Companies will form universal charging stations to charge their own electrical vehicles or to charge the vehicles of other companies, thus minimizing the charging time that is spent on battery charging by installing battery exchange units or battery rental stations. Thus, the world can shift to an era of smarter transportation and wiser energy management.

In the present study, V2G and V2H technologies were introduced, and information on their structures and components was given. The charging system of the batteries in the vehicles was narrated with brief talks about battery life, its health indicators, how different factors affect its health, and battery management systems. In addition, the topic was enhanced by including the power electronics topologies that are widely used in this sector. The basic idea and circuit structure were assessed for each converter topology. The communication standards used in this technology were indicated. A comparative assessment of the overall performance of different types of AC/DC and DC/DC converters was made that includes several parameters. Finally, a SWOT analysis of the V2G technology was made to elucidate that this technology will eventually bring in good results for the energy sector.

Author Contributions: All authors contributed equally to the research activities and for its final presentation as a full manuscript.

Funding: No source of funding was attained for this research activity.

Acknowledgments: The authors would like to express their sincere gratitude to Oregon Renewable Energy Center (OREC) for their assistance with the experimentation required during the course of this research. Their contributions have greatly improved the content of the manuscript.

Conflicts of Interest: The authors declare no conflict of interest.

References

1. Ota, Y.; Taniguchi, H.; Baba, J.; Yokoyama, A. Implementation of autonomous distributed V2G to electric vehicle and DC charging system. *Electr. Power Syst. Res.* **2015**, *120*, 177–183. [CrossRef]
2. Winkler-Goldstein, R.; Rastetter, A. Power to Gas: The Final Breakthrough for the Hydrogen Economy? *Green* **2013**, *3*. [CrossRef]
3. Xueqin, L.; Fuzhen, H.; Gang, L.; Rongfu, Q. The challenges of technologies for fuel cell and its application on vehicles. In Proceedings of the 2009 IEEE 6th International Power Electronics and Motion Control Conference, Wuhan, China, 17–20 May 2009; p. 2328. [CrossRef]
4. Bossel, U. Does a Hydrogen Economy Make Sense? *Proc. IEEE* **2006**, *94*, 1826–1837. [CrossRef]
5. White, C.D.; Zhang, K.M. Using Vehicle-to-Grid Technology for Frequency Regulation and Peak-Load Reduction. *J. Power Sources* **2011**, *196*, 3972–3979. [CrossRef]
6. Kisacikoglu, M.C.; Ozpineci, B.; Tolbert, L.M. EV/PHEV Bidirectional Charger Assessment for V2G Reactive Power Operation. *IEEE Trans. Power Electron.* **2013**, *28*, 5717–5727. [CrossRef]
7. De Freige, M.; Ross, M.; Joos, G.; Dubois, M. Power & Energy Ratings Optimization in a Fast-Charging Station for PHEV Batteries. In Proceedings of the 2011 IEEE International Electric Machines & Drives Conference (IEMDC), Victoria, BC, Canada, 24–26 August 2011; p. 486. [CrossRef]
8. Pearre, N.S.; Ribberink, H. Review of research on V2X technologies, strategies, and operations. *Renew. Sustain. Energy Rev.* **2019**, *105*, 61–70. [CrossRef]
9. Arancibia, A.; Strunz, K. Modeling of an Electric Vehicle Charging Station for Fast DC Charging. In Proceedings of the 2012 IEEE International Electric Vehicle Conference, Greenville, SC, USA, 4–8 March 2012; pp. 1–6. [CrossRef]
10. Yilmaz, M.; Krein, P.T. Review of Battery Charger Topologies, Charging Power Levels, and Infrastructure for Plug-In Electric and Hybrid Vehicles. *IEEE Trans. Power Electron.* **2013**, *28*, 2151–2169. [CrossRef]
11. Musavi, F.; Edington, M.; Eberle, W.; Dunford, W.G. Energy Efficiency in Plug-in Hybrid Electric Vehicle Chargers: Evaluation and Comparison of Front end AC-DC Topologies. In Proceedings of the 2011 IEEE Energy Conversion Congress and Exposition, Phoenix, AZ, USA, 17–22 September 2011; pp. 273–278. [CrossRef]
12. Kwon, M.; Jung, S.; Choi, S. A high efficiency bi-directional EV charger with seamless mode transfer for V2G and V2H application. In Proceedings of the 2015 IEEE Energy Conversion Congress and Exposition (ECCE), Montreal, QC, Canada, 20–24 September 2015; pp. 5394–5396. [CrossRef]
13. Rahman, I.; Vasant, P.M.; Singh BS, M.; Abdullah-Al-Wadud, M.; Adnan, N. Review of Recent Trends in Optimization Techniques for Plug-in Hybrid, And Electric Vehicle Charging Infrastructures. *Renew. Sustain. Energy Rev.* **2016**, *58*, 1039–1047. [CrossRef]
14. Colak, I.; Kabalci, E.; Bayindir, R. Review of multilevel voltage source inverter topologies and control schemes. *Energy Convers. Manag.* **2011**, *52*, 1114–1115. [CrossRef]
15. Colak, I.; Kabalci, E.; Fulli, G.; Lazarou, S. A survey on the contributions of power electronics to smart grid systems. *Renew. Sustain. Energy Rev.* **2015**, *47*, 562–579. [CrossRef]
16. Fathabadi, H. Utilization of electric vehicles and renewable energy sources used as distributed generators for improving characteristics of electric power distribution systems. *Energy* **2015**, *90*, 1100–1111. [CrossRef]
17. Erb, D.C.; Onar, O.C.; Khaligh, A. Bi-Directional Charging Topologies for Plug-In Hybrid Electric Vehicles. In Proceedings of the 2010 Twenty-Fifth Annual IEEE Applied Power Electronics Conference and Exposition (APEC), Palm Springs, CA, USA, 21–25 February 2010; pp. 2066–2067. [CrossRef]
18. Jiang, J.; Bao, Y.; Wang, L. Topology of a Bidirectional Converter for Energy Interaction between Electric Vehicles and the Grid. *Energies* **2014**, *7*, 4858. [CrossRef]

19. Gao, S.K.T.C.; Liu, C.; Wu, D. Optimal Control Framework and Scheme for Integrating Plug-in Hybrid Electric Vehicles into Grid. *J. Asian Electr. Veh.* **2011**, *9*, 1473–1479. [CrossRef]

20. Pinto, J.G.; Monteiro, V.; Gonçalves, H.; Exposto, B.; Pedrosa, D.; Couto, C.; Afonso, J.L. Bidirectional battery charger with Grid-to-Vehicle, Vehicle-to-Grid and Vehicle-to-Home technologies. In Proceedings of the IECON 2013—39th Annual Conference of the IEEE Industrial Electronics Society, Vienna, Austria, 10–13 November 2013; pp. 5934–5936. [CrossRef]

21. Narula, A.; Verma, V. Bi-Directional Trans-Z Source Boost Converter for G2V/V2G Applications. In Proceedings of the 2017 IEEE Transportation Electrification Conference (ITEC-India), Pune, India, 13–16 December 2017; pp. 1–6. [CrossRef]

22. Liu, H.; Hu, Z.; Song, Y.; Wang, J.; Xie, X. Vehicle-to-Grid Control for Supplementary Frequency Regulation Considering Charging Demands. *IEEE Trans. Power Syst.* **2015**, *30*, 3110. [CrossRef]

23. Hota, A.R.; Juvvanapudi, M.; Bajpai, P. Issues and solution approaches in PHEV integration to the smart grid. *Renew. Sustain. Energy Rev.* **2014**, *30*, 217–229. [CrossRef]

24. Bahrami, S.; Wong, V.W. A potential game framework for charging PHEVs in smart grid. In Proceedings of the 2015 IEEE Pacific Rim Conference on Communications, Computers and Signal Processing (PACRIM), Victoria, BC, Canada, 24–26 August 2015; p. 28. [CrossRef]

25. Harighi, T.; Bayindir, R.; Padmanaban, S.; Mihet-Popa, L.; Hossain, E. An Overview of Energy Scenarios, Storage Systems and the Infrastructure for Vehicle-to-Grid Technology. *Energies* **2018**, *11*, 2174. [CrossRef]

26. Colak, I.; Fulli, G.; Sagiroglu, S.; Yesilbudak, M.; Covrig, C.F. Smart grid projects in Europe: Current status, maturity and future scenarios. *Appl. Energy* **2015**, *152*, 58–70. [CrossRef]

27. Colak, I.; Sagiroglu, S.; Fulli, G.; Yesilbudak, M.; Covrig, C.F. A Survey on the Critical Issues in Smart Grid Technologies. *Renew. Sustain. Energy Rev.* **2016**, *54*, 396–405. [CrossRef]

28. Bayindir, R.; Colak, I.; Fulli, G.; Demirtas, K. Demirtas K. Smart grid technologies and applications. *Renew. Sustain. Energy Rev.* **2016**, *66*, 499–516. [CrossRef]

29. Ansari, K.; Feng, Y. Design of an Integration Platform for V2X Wireless Communications and Positioning Supporting C-ITS Safety Applications. *J. Glob. Position. Syst.* **2013**, *12*, 38–52. [CrossRef]

30. Ghods, A.; Severi, S.; Abreu, G. Localization in V2X communication networks. In Proceedings of the 2016 IEEE Intelligent Vehicles Symposium (IV), Gothenburg, Sweden, 19–22 June 2016; pp. 5–9. [CrossRef]

31. Siegel, J.E. *CloudThink and the Avacar: Embedded Design to Create Virtual Vehicles for Cloud-Based Informatics, Telematics, and Infotainment*; Cambridge Massachusetts Inst. Technol: Cambridge, MA, USA, 2013.

32. Wilhelm, E.; Siegel, J.; Mayer, S.; Sadamori, L.; Dsouza, S.; Chau, C.K.; Sarma, S. Cloudthink: A scalable secure platform for mirroring transportation systems in the cloud. *Transport* **2015**, *30*, 320–329. [CrossRef]

33. Rinaldi, S.; Pasetti, M.; Trioni, M.; Vivacqua, G. On the Integration of E-Vehicle Data for Advanced Management of Private Electrical Charging Systems. In Proceedings of the 2017 IEEE International Instrumentation and Measurement Technology Conference (I2MTC), Turin, Italy, 22–25 May 2017; pp. 1–6. [CrossRef]

34. Siegel, J.E.; Erb, D.C.; Sarma, S.E. A Survey of the Connected Vehicle Landscape—Architectures, Enabling Technologies, Applications, and Development Areas. *IEEE Trans. Intell. Transp. Syst.* **2018**, *19*, 2391–2406. [CrossRef]

35. Liu, H.; Yang, Y.; Qi, J.; Li, J.; Wei, H.; Li, P. Frequency droop control with scheduled charging of electric vehicles. *IET Gener. Transm. Distrib.* **2016**, *11*, 649–656. [CrossRef]

36. Orihara, D.; Kimura, S.; Saitoh, H. Frequency Regulation by Decentralized V2G Control with Consensus-Based SOC Synchronization. *IFAC-PapersOnLine* **2018**, *51*, 604–606. [CrossRef]

37. Tuttle, D.P.; Fares, R.L.; Baldick, R.; Webber, M.E. Plug-In Vehicle to Home (V2H) duration and power output capability. In Proceedings of the 2013 IEEE Transportation Electrification Conference and Expo (ITEC), Detroit, MI, USA, 16–19 June 2013; pp. 1–7. [CrossRef]

38. Han, S.; Han, S.; Sezaki, K. Estimation of achievable power capacity from plug-in electric vehicles for V2G frequency regulation: Case studies for market participation. *IEEE Trans. Smart Grid* **2011**, *2*, 632–641. [CrossRef]

39. Liu, Y.J.C.T.; Chen, H.W.; Chang, T.K.; Lan, P.H. Power quality measurements of low-voltage distribution system with smart electric vehicle charging infrastructures. In Proceedings of the 2014 16th International Conference on Harmonics and Quality of Power (ICHQP), Bucharest, Romania, 25–28 May 2014; pp. 631–635. [CrossRef]

40. Berthold, F.; Ravey, A.; Blunier, B.; Bouquain, D.; Williamson, S.; Miraoui, A. Design and Development of a Smart Control Strategy for Plug-In Hybrid Vehicles Including Vehicle-to-Home Functionality. *IEEE Trans. Transp. Electrif.* **2015**, *1*, 168–177. [CrossRef]

41. Weiller, C. Plug-in hybrid electric vehicle impacts on hourly electricity demand in the United States. *Energy Policy* **2011**, *39*, 3766–3778. [CrossRef]

42. Liu, C.; Chau, K.T.; Wu, D.; Gao, S. Opportunities and challenges of vehicle-to-home, vehicle-to-vehicle, and vehicle-to-grid technologies. *Proc. IEEE* **2013**, *101*, 2409–2419. [CrossRef]

43. Ahmad, F.; Alam, M.S.; Asaad, M. Developments in EVs charging infrastructure and energy management system for smart microgrids including EVs. *Sustain. Cities Soc.* **2017**, *35*, 552–564. [CrossRef]

44. Mwasilu, F.; Justo, J.J.; Kim, E.K.; Do, T.D.; Jung, J.W. Electric vehicles and smart grid interaction: A review on vehicle to grid and renewable energy sources integration. *Renew. Sustain. Energy Rev.* **2014**, *34*, 501–516. [CrossRef]

45. Galus, M.D.; Koch, S.; Andersson, G. Provision of load frequency control by PHEVS, controllable loads, and a cogeneration unit. *IEEE Trans. Ind. Electron.* **2011**, *58*, 4568–4582. [CrossRef]

46. Un-Noor, F.; Padmanaban, S.; Mihet-Popa, L.; Mollah, N.M.; Hossain, E. A Comprehensive Study of Key Electric Vehicle (EV) Components, Technologies, Challenges, Impacts, and Future Direction of Development. *Energies* **2017**, *10*, 1217. [CrossRef]

47. Ota, Y.; Taniguchi, H.; Nakajima, T.; Liyanage, K.M.; Baba, J.; Yokoyama, A. Autonomous distributed V2G (Vehicle-to-Grid) satisfying scheduled charging. *IEEE Trans. Smart Grid* **2012**, *3*, 559–564. [CrossRef]

48. Shimizu, K.; Masuta, T.; Ota, Y.; Yokoyama, A. A New Load Frequency Control Method in Power System Using Vehicle-To-Grid System Considering Users Convenience. In Proceedings of the 17th Power System Computation Conference, Stockholm, Sweden, 22–26 August 2011; pp. 1–7.

49. Ferreira, R.J.; Miranda, L.M.; Araújo, R.E.; Lopes. J.P. A New Bi-Directional Charger For Vehicle-To-Grid Integration. In Proceedings of the 2011 2nd IEEE PES International Conference and Exhibition on Innovative Smart Grid Technologies, Manchester, UK, 5–7 December 2011; pp. 1–5. [CrossRef]

50. Hossain, E.; Murtaugh, D.; Mody, J.; Faruque, H.M.; Sunny, M.S.; Mohammad, N. A Comprehensive Review on Second-Life Batteries: Current State, Manufacturing Considerations, Applications, Impacts, Barriers & Potential Solutions, Business Strategies, and Policies. *IEEE Access* **2019**, *7*, 73215–73252. [CrossRef]

51. McLoughlin, F.; Conlon, M. *Secondary Re-Use of Batteries from Electric Vehicles for Building Integrated Photo-Voltaic (BIPV) Applications*; Dublin Institute of Technology: Dublin, Ireland, 2015.

52. Andersen, P.B.; Garcia-Valle, R.; Kempton, W. A Comparison of Electric Vehicle Integration Projects. In Proceedings of the 2012 3rd IEEE PES Innovative Smart Grid Technologies Europe (ISGT Europe), Berlin, Germany, 14–17 October 2012; pp. 1–7. [CrossRef]

53. Su, G.J.; Tang, L. Using Onboard Electrical Propulsion System to Provide Plug-incharging, V2G and Mobile Power Generation Capabilities for Hevs. In Proceedings of the IEEE Electric Vehicle Conference, Greenville, SC, USA, 4–8 March 2012; pp. 1–8. [CrossRef]

54. Madawala, U.K.; Thrimawithana, D.J. A Bidirectional Inductive Power Interface for Electric Vehicles in V2G Systems. *IEEE Trans. Ind. Electron.* **2011**, *58*, 4789–4796. [CrossRef]

55. Nissan Leaf—Battery Capacity Loss. Available online: http://www.electricvehiclewiki.com/wiki/battery-capacity-loss/ (accessed on 17 September 2019).

56. Harighi, T.; Padmanaban, S.; Bayindir, R.; Hossain, E.; Holm-Nielsen, J.B. Electric Vehicle Charge Stations Location Analysis and Determination—Ankara (Turkey) Case Study. *Energies* **2019**, *12*, 3472. [CrossRef]

57. Heydarian-Forushani, E.; Golshan, M.; Shafie-khah, M. Flexible interaction of plug-in electric vehicle parking lots for efficient wind integration. *Appl. Energy* **2016**, *179*, 338–349. [CrossRef]

58. He, Y.; Kockelman, K.M.; Perrine, K.A. Optimal locations of U.S. fast charging stations for long-distance trip completion by battery electric vehicles. *J. Clean. Prod.* **2019**, *214*, 452–461. [CrossRef]

59. Zarandi, M.F.; Davari, S.; Sisakht, S.H. The large scale maximal covering location problem. *Sci. Iran.* **2011**, *18*, 1564–1570. [CrossRef]

60. Cui, S.; Zhao, H.; Zhang, C. Locating Charging Stations of Various Sizes with Different Numbers of Chargers for Battery Electric Vehicles. *Energies* **2018**, *11*, 3056. [CrossRef]

61. Said, D.; Cherkaoui, S.; Khoukhi, L. Queuing model for EVs charging at public supply stations. In Proceedings of the 9th International Wireless Communications and Mobile Computing Conference (IWCMC), Cagliari, Italy, 1–5 July 2013; pp. 65–70. [CrossRef]

62. Said, D.; Cherkaoui, S.; Khoukhi, L. Multi-priority queuing for electric vehicles charging at public supply stations with price variation: Multi-priority queuing EV charging with price variation. *Wirel. Commun. Mob. Comput.* **2014**, *15*. [CrossRef]

63. He, F.; Yin, Y.; Lawphongpanich, S. Network equilibrium models with battery electric vehicles. *Transp. Res. Part. B Methodol.* **2014**, *67*, 306–319. [CrossRef]

64. Habib, S.; Kamran, M.; Rashid, U. Impact analysis of vehicle-to-grid technology and charging strategies of electric vehicles on distribution networks—A review. *J. Power Sources* **2015**, *277*, 205–214. [CrossRef]

65. Liu, G.; Kang, L.; Luan, Z.; Qiu, J.; Zheng, F. Charging Station and Power Network Planning for Integrated Electric Vehicles (EVs). *Energies* **2019**, *12*, 2595. [CrossRef]

66. Aveklouris, A.; Nakahira, Y.; Vlasiou, M.; Zwart, B. Electric vehicle charging: A queueing approach. *ACM SIGMETRICS Perform. Eval. Rev.* **2017**, *45*, 33–35. [CrossRef]

67. Deveci, F.; Boztepe, M. Design of Temperature Compensated Charger for Lead-Acid Battery. In Proceedings of the National Conference on Electrical, Electronics and Computer Engineering, Bursa, Turkey, 2–5 December 2010; pp. 269–272.

68. Bhatti, A.R.; Salam, Z.; Aziz MJ, B.A.; Yee, K.P.; Ashique, R.H. Electric vehicles charging using photovoltaic: Status and technological review. *Renew. Sustain. Energy Rev.* **2016**, *54*, 34–47. [CrossRef]

69. Li, S.; Zhang, C.; Xie, S. Research on Fast Charge Method for Lead-Acid Electric Vehicle Batteries. In Proceedings of the International Workshop on Intelligent Systems and Applications, Wuhan, China, 23–24 May 2009; pp. 1–5.

70. Huet, F. A review of impedance measurements for determination of the state-of-charge or state-of-health of secondary batteries. *J. Power Sources* **1998**, *70*, 56–69. [CrossRef]

71. Huria, T.; Ceraolo, M.; Gazzarri, J.; Jackey, R. Simplified Extended Kalman Filter Observer for SOC Estimation of Commercial Power-Oriented LFP Lithium Battery Cells. *SAE Tech. Pap.* **2013**, 1–10. [CrossRef]

72. Andre, D.; Meiler, M.; Steiner, K.; Wimmer, C.; Soczka-Guth, T.; Sauer, D.U. Characterization of high-power lithium-ion batteries by electrochemical impedance spectroscopy. I. Experimental investigation. *J. Power Sources* **2011**, *196*, 5334–5338. [CrossRef]

73. Wei, X.; Zhu, B.; Xu, W. Internal Resistance Identification in Vehicle Power Lithium-Ion Battery and Application in Lifetime Evaluation. In Proceedings of the International Conference on In Measuring Technology and Mechatronics Automation, Zhangjiajie, China, 11–12 April 2009; pp. 388–392. [CrossRef]

74. Zhang, Y.; Wang, C.Y.; Tang, X. Cycling Degradation of an Automotive LiFePO4 Lithium-Ion Battery. *J. Power Sources* **2011**, *196*, 1513–1518. [CrossRef]

75. Shim, J.; Kostecki, R.; Richardson, T.; Song, X.; Striebel, K.A. Electrochemical Analysis for Cycle Performance and Capacity Fading of a Lithium-Ion Battery Cycled at Elevated Temperature. *J. Power Sources* **2012**, *112*, 220–229. [CrossRef]

76. Peterson, S.B.; Apt, J.; Whitacre, J.F. Lithium-Ion Battery Cell Degradation Resulting from Realistic Vehicle and Vehicle-to-Grid Utilization. *J. Power Sources* **2010**, *195*, 2385–2388. [CrossRef]

77. Hossain, E.; Perez, R.; Nasiri, A.; Bayindir, R. Stability improvement of microgrids in the presence of constant power loads. *Int. J. Electr. Power Energy Syst.* **2018**, *96*, 442–456. [CrossRef]

78. Reddy, B.M.; Samuel, P. A Comparative Analysis of Non-Isolated Bi-directional DC-DC Converters. In Proceedings of the 1st IEEE International Conference on Power Electronics, Intelligent Control. and Energy Systems (ICPEICES-2016), Delhi, India, 4–6 July 2016; pp. 1–6.

79. Elankurisil, S.A.; Dash, S. Comparison of Isolated and Non-isolated Bi-directional DC-DC Converters. *J. Eng.* **2011**, *21*, 2341–2347.

80. Mehdipour A, F.S. Comparison of Three Isolated Bi-Directional DC/DC Converter Topologies for Backup Photovoltaic Application. In Proceedings of the 2nd International Conference on Electric Power and Energy Conversion Systems (EPECS), Sharjah, United Arab Emirates, 15–17 November 2011; pp. 1–5.

81. Xu, Y.C.Y.; Huang, A.Q. Five Level Bi-Directional Converter for Renewable Energy Generation. In Proceedings of the IECON 2014—40th Annual Conference of the IEEE Industrial Electronics Society, Dallas, TX, USA, 29 October–1 November 2014; pp. 5514–5516.

82. Rei, R.J.; Soares, F.J.; Almeida, P.R.; Lopes, J.P. Grid Interactive Charging Control for Plug-in Electric Vehicles. In Proceedings of the 13th International IEEE Conference on Intelligent Transportation Systems, Funchal, Portugal, 19–22 September 2010; p. 386.

83. Freire, R.; Delgado, J.; Santos, J.M.; De Almeida, A.T. Integration of Renewable Energy Generation with EV Charging Strategies to Optimize Grid Load Balancing. In Proceedings of the 13th International IEEE Conference on Intelligent Transportation Systems, Funchal, Portugal, 19–22 September 2010; pp. 392–395.
84. Ferreira, J.C.; Trigo, P.; da Silva, A.R.; Coelho, H.; Afonso, J.L. Simulation of Electrical Distributed Energy Resources for Electrical Vehicles Charging Process Strategy. In Proceedings of the Second Brazilian Workshop on Social Simulation (BWSS 2010), Washington, DC, USA, 24–25 October 2010; pp. 82–89.
85. Hernández, S.S.; Galindo, P.P.; López, A.Q. Technology to integrate EV inside smart grids. In Proceedings of the 2010 7th International Conference on the European Energy Market, Madrid, Spain, 23–25 June 2010; pp. 1–6.
86. Jarnut, M.; Benysek, G. Application of Power Electronics Devices in Smart Grid and V2G(Vehicle to Grid) technologies. *Prz. Elektrotech.* **2010**, *86*, 93–94.
87. Kisacikoglu, M.C.; Ozpineci, B.; Tolbert, L.M. Effects of V2G Reactive Power Compensation on the Component Selection in an EV or PHEV Bidirectional Charger. In Proceedings of the IEEE Energy Conversion Congress and Exposition, Atlanta, GA, USA, 12–16 September 2010; pp. 870–877.
88. Pinto, J.G.; Monteiro, V.; Gonçalves, H.; Afonso, J.L. Onboard reconfigurable battery charger for electric vehicles with traction-to-auxiliary mode. *IEEE Trans. Veh. Technol.* **2014**, *63*, 1104–1113. [CrossRef]
89. Khan, M.A.; Husain, I.; Sozer, Y. Bi-directional DC-DC converter with overlapping input and output voltage ranges and vehicle to grid energy transfer capability. In Proceedings of the 2012 IEEE International Electric Vehicle Conference (IEVC), Greenville, SC, USA, 4–8 March 2012; pp. 1–7.
90. Das, P.; Pahlevaninezhad, M.; Moschopoulos, G. Analysis and Design of a New AC–DC Single-Stage Full-Bridge PWM Converter with Two Controllers. *IEEE Trans. Ind. Electron.* **2013**, *60*, 4930–4946. [CrossRef]
91. Wijeratne, D.S.; Moschopoulos, G. A Comparative Study of Two Buck-Type Three-Phase Single-Stage AC–DC Full-Bridge Converters. *IEEE Trans. Power Electron.* **2014**, *29*, 1632–1645. [CrossRef]
92. Xie, Y.; Sun, J.; Freudenberg, J.S. Power flow characterization of a bidirectional galvanically isolated high-power DC/DC converter over a wide operating range. *IEEE Trans. Power Electron.* **2010**, *25*, 54–66.
93. Shi, X.; Jiang, J.; Guo, X. An efficiency-optimized isolated bidirectional DC-DC converter with extended power range for energy storage systems in microgrids. *Energies* **2013**, *6*, 27–44. [CrossRef]
94. Kanaan, H.Y.; Caron, M.; Al-Haddad, K. Design and implementation of a two-stage grid-connected high efficiency power load emulator. *IEEE Trans. Power Electron.* **2014**, *29*, 3997–4006. [CrossRef]
95. Zhao, B.; Song, Q.; Liu, W.; Sun, Y. Overview of dual-active bridge isolated bidirectional dc-dc converter for high frequency-link power-conversion system. *IEEE Trans. Power Electron.* **2014**, *29*, 4091–4106. [CrossRef]
96. Lu, L.; Han, X.; Li, J.; Hua, J.; Ouyang, M. A review on the key issues for lithium-ion battery management in electric vehicles. *J. Power Sources* **2013**, *226*, 272–288. [CrossRef]
97. He, Y.; Venkatesh, B.; Guan, L. Optimal scheduling for charging and discharging of electric vehicles. *IEEE Trans. Smart Grid* **2012**, *3*, 1095–1105. [CrossRef]
98. Wei, Z.; Li, Y.; Zhang, Y.; Cai, L. Intelligent parking garage EV charging scheduling considering battery charging characteristic. *IEEE Trans. Ind. Electron.* **2018**, *65*, 2806–2816. [CrossRef]
99. Mahmud, K.; Town, G.E.; Morsalin, S.; Hossain, M.J. Integration of electric vehicles and management in the internet of energy. *Renew. Sustain. Energy Rev.* **2018**, *82*, 4179–4203. [CrossRef]
100. Saltanovs, R.; Krivchenkov, A.; Krainyukov, A. Analysis of effective wireless communications for V2G applications and mobile objects. In Proceedings of the 58th International Scientific Conference on Power and Electrical Engineering of Riga Technical University (RTUCON), Riga, Latvia, 12–13 October 2017.
101. Du, Y.; Lukic, S.; Jacobson, B.; Huang, A. Review of high power isolated bi-directional DC-DC converters for PHEV/EV DC charging infrastructure. In Proceedings of the 2011 IEEE Energy Conversion Congress and Exposition, Phoenix, AZ, USA, 16–21 September 2011; pp. 553–558.
102. Guille, C.; Gross, G. A conceptual framework for the vehicle-to-grid (V2G) implementation. *Energy Policy* **2009**, *37*, 4379–4390. [CrossRef]
103. Rezaee, S.; Farjah, E. A DC–DC Multiport Module for Integrating Plug-In Electric Vehicles in a Parking Lot: Topology and Operation. *IEEE Trans. Power Electron.* **2014**, *29*, 5688–5695. [CrossRef]
104. Richardson, D.B. Electric Vehicles and the Electric Grid: A Review of Modeling Approaches, Impacts, and Renewable Energy Integration. *Renew. Sustain. Energy Rev.* **2013**, *19*, 247–248. [CrossRef]
105. Tuttle, D.P.; Baldick, R. The evolution of plug-in electric vehicle-grid interactions. *IEEE Trans. Smart Grid* **2012**, *3*, 500–506. [CrossRef]

106. Guo, H.; Wu, Y.; Bao, F.; Chen, H.; Ma, M. A unique batch authentication protocol for vehicle-to-grid communications. *IEEE Trans. Smart Grid* **2011**, *2*, 707–708. [CrossRef]

107. Ehsani, M.; Falahi, M.; Lotfifard, S. Vehicle to grid services: Potential and applications. *Energies* **2012**, *5*, 4076–4090. [CrossRef]

108. Keyhani, H.; Toliyat, H.A. A ZVS single-inductor multi-input multi output DC-DC converter with the step up/down capability. In Proceedings of the 2013 IEEE Energy Conversion Congress and Exposition, Denver, CO, USA, 15–19 September 2013; pp. 5546–5547.

109. Schmutzler, J.; Gröning, S.; Wietfeld, C. Management of Distributed Energy Resources in IEC 61850 using Web Services on Devices. In Proceedings of the 2011 IEEE International Conference on Smart Grid Communications (SmartGridComm), Brussels, Belgium, 17–20 October 2011; pp. 315–316.

110. Schmutzler, J.; Wietfeld, C.; Jundel, S.; Voit, S. A Mutual Charge Schedule Information Model for the Vehicle-to-Grid Communication Interface. In Proceedings of the 2011 IEEE Vehicle Power and Propulsion Conference, Chicago, IL, USA, 5–8 September 2011; pp. 1–6.

111. Kiokes, G.; Zountouridou, E.; Papadimitriou, C.; Dimeas, A.; Hatziargyriou, N. Papadimitriou, A. In Dimeas, N. Hatziargyriou. Development of an Integrated Wireless Communication System for Connecting Electric Vehicles to the Power Grid. In Proceedings of the 2015 International Symposium on Smart Electric Distribution Systems and Technologies (EDST), Vienna, Austria, 8–11 September 2015; pp. 296–301.

112. Hu, J.; Morais, H.; Sousa, T.; Lind, M. Electric vehicle fleet management in smart grids: A review of services, optimization and control aspects. *Renew. Sustain. Energy Rev.* **2016**, *56*, 1207–1226. [CrossRef]

113. Bessa, R.J.; Matos, M.A. Economic and technical management of an aggregation agent for electric vehicles: A literature survey. *Eur. Trans. Electr. Power* **2012**, *22*, 334–350. [CrossRef]

114. Ustun, T.S.; Ozansoy, C.R.; Zayegh, A. Implementing vehicle-to-grid (V2G) technology with IEC 61850-7-420. *IEEE Trans. Smart Grid* **2013**, *4*, 1180–1187. [CrossRef]

115. Käbisch, S.; Schmitt, A.; Winter, M.; Heuer, J. Interconnections and Communications of Electric Vehicles and Smart Grids. In Proceedings of the 2010 First IEEE International Conference on Smart Grid Communications, Gaithersburg, MD, USA, 4–6 October 2010; pp. 161–166.

116. Liu, Y.; Mitchem, S.C. Implementation of V2G Technology Using DC Fast Charging. In Proceedings of the 2013 International Conference on Connected Vehicles and Expo (ICCVE), Las Vegas, NV, USA, 2–6 December 2013; pp. 734–735.

117. Han, H.; Liu, Y.; Sun, Y.; Wang, H.; Su, M. A Single-Phase Current-Source Bidirectional Converter For V2G Applications. *J. Power Electron.* **2014**, *14*, 458–467. [CrossRef]

118. Su, M.; Li, H.; Sun, Y.; Xiong, W. A High-Efficiency Bidirectional Ac/Dc Topology for V2G Applications. *Power Electron.* **2014**, *14*, 899–907. [CrossRef]

119. Zhao, T.; Li, Y.; Pan, X.; Wang, P.; Zhang, J. Real-time optimal energy and reserve management of electric vehicle fast charging station: Hierarchical game approach. *IEEE Trans. Smart Grid* **2017**, *99*, 1–9. [CrossRef]

120. Verma, A.K.; Singh, B.; Shahani, D.T. Grid to Vehicle and Vehicle to Grid Energy Transfer using Single-Phase Bidirectional AC-DC Converter and Bidirectional DC-DC Converter. In Proceedings of the 2011 International Conference on Energy, Automation and Signal, Bhubaneswa, India, 28–30 December 2011; pp. 1–5.

121. Pahlevani, M.; Jain, P. A Fast DC-Bus Voltage Controller for Bidirectional Single Phase AC/DC Converters. *IEEE Trans. Power Electron.* **2015**, *30*, 4536–4547. [CrossRef]

122. Peng, T.; Yang, P.; Dan, H.; Wang, H.; Han, H.; Yang, J.; Wheeler, P. A single-phase bidirectional AC/DC Converter for V2G applications. *Energies* **2017**, *10*, 881. [CrossRef]

123. Choi, W.; Han, D.; Morris, C.T.; Sarlioglu, B. Achieving high efficiency using SiC MOSFETs and reduced output filter for grid-connected V2G Inverter. In Proceedings of the IECON 2015—41st Annual Conference of the IEEE Industrial Electronics Society, Yokohama, Japan, 9–12 November 2015; Volume 201, pp. 3052–3056.

124. Onar, O.C.; Kobayashi, J.; Erb, D.C.; Khaligh, A. A Bidirectional High-Power-Quality Grid Interface with a Novel Bidirectional Non-Inverted Buck–Boost Converter for PHEVS. *IEEE Trans. Veh. Technol.* **2012**, *61*, 2018–2032. [CrossRef]

125. Jauch, F.; Biela, J. Single-phase single-stage bidirectional isolated ZVS AC-DC converter with PFC. In Proceedings of the 2012 15th International Power Electronics and Motion Control Conference (EPE/PEMC), Novi Sad, Serbia, 4–6 September 2012; Volume 5, pp. 1–8.

126. Sun, Y.; Liu, W.; Su, M.; Li, X.; Wang, H.; Yang, J. A Unified Modeling and Control of a Multi-Functional Current Source-Typed Converter for V2G Application. *Elect. Power Syst. Res.* **2014**, *106*, 12–19. [CrossRef]
127. Hegazy, O.; Van Mierlo, J.; Lataire, P. Control and Analysis of An Integrated Bidirectional DC/AC and DC/DC Converters for Plug-In Hybrid Electric Vehicle Applications. *J. Power Electron.* **2011**, *11*, 408–410. [CrossRef]

MDPI

St. Alban-Anlage 66

4052 Basel

Switzerland

Tel. +41 61 683 77 34

Fax +41 61 302 89 18

www.mdpi.com

Energies Editorial Office

E-mail: energies@mdpi.com

www.mdpi.com/journal/energies